Mathematics, Magic and Mystery

MARTIN GARDNER

DOVER PUBLICATIONS, INC., NEW YORK

Published in Canada by General Publishing Company, Ltd., 30 Lesmill Road, Don Mills, Toronto, Ontario.
Published in the United Kingdom by Constable and Company, Ltd.

Mathematics, Magic and Mystery is a new work, published for the first time by Dover Publications, Inc., in 1956.

Standard Book Number: 486-20335-2
Library of Congress Catalog Card Number: 57-1546

Manufactured in the United States of America
Dover Publications, Inc.
180 Varick Street
New York, N. Y. 10014

Table of Contents

v

Preface

Like many another hybrid subject matter, mathematical magic is often viewed with a double disdain. Mathematicians are inclined to regard it as trivial play, magicians to dismiss it as dull magic. Its practitioners, to rephrase an epigram about biophysicists, are apt to bore their mathematical friends with talk of magic, their magic friends with talk of mathematics, and each other with talk of politics. There is truth in all these animadversions. Mathematical magic is not—let us face it— the sort of magic with which one can hold spellbound an audience of nonmathematically minded people. Its tricks take too long and they have too little dramatic effect. Nor is one likely to obtain profound mathematical insights from contemplating tricks of a mathematical character.

Nevertheless, mathematical magic, like chess, has its own curious charms. Chess combines the beauty of mathematical structure with the recreational delights of a competitive game. Mathematical magic combines the beauty of mathematical structure with the entertainment value of a trick. It is not surprising, therefore, that the delights of mathematical magic are greatest for those who enjoy both conjuring and mathematical recreations.

W. W. Rouse Ball (1851–1925), a fellow in mathematics at Trinity College, Cambridge, and author of the well known MATHEMATICAL RECREATIONS AND ESSAYS, was such an individual. Throughout his life, he took an active interest in legerdemain. He founded and was the first president of The Pentacle Club, a Cambridge University magic society that continues to flourish today. His classic reference work contains many early examples of mathematical conjuring.

So far as I am aware, the chapters to follow represent the first attempt to survey the entire field of modern mathematical

magic. Most of the material has been drawn from the literature of conjuring and from personal contacts with amateur
and professional magicians rather than from the literature of
mathematical recreations. It is the magician, not the mathematician, who has been the most prolific in creating mathematical tricks during the past half-century. For this reason,
students of recreational mathematics not familiar with modern
conjuring are likely to find here a rich new field—a field of
which they may well have been totally unaware.

It is a field in its infancy. It is a field in which dozens of
startling new effects may be invented before this book has been
a year on sale. Because its principles can be grasped quickly,
without training in higher mathematics, perhaps you the
reader may play some part in the rapid growth of this odd and
delightful pastime.

I want to thank Professor Jekuthiel Ginsburg, editor of
Scripta Mathematica, for permission to reprint material from
four articles which I contributed to his excellent journal.
Paul Curry, Stewart James, Mel Stover, and N. T. Gridgeman
gave generously of their time and knowledge in reading the
manuscript, correcting mistakes, and offering valuable sugges-
tions. Other friends who supplied material and information
are too numerous to mention. Finally, I owe a special debt
to my wife for uninhibited and indispensable criticism, as well
as tireless assistance in all phases of the book's preparation.

                                                      MARTIN GARDNER
New York, N.Y., 1955

Chapter One

Tricks with Cards—Part One

Playing cards possess five basic features that can be exploited in devising card tricks of a mathematical character.

(1) They can be used as counting units, without reference to their faces, as one might use pebbles, matches, or pieces of paper.

(2) The faces have numerical values from one to thirteen (considering the jack, queen, and king as 11, 12, and 13 respectively).

(3) They are divided into four suits, as well as into red and black cards.

(4) Each card has a front and back.

(5) Their compactness and uniform size make it easy to arrange them into various types of series and sets, and conversely, arrangements can be destroyed quickly by shuffling.

Because of this richness of appropriate properties, mathematical card tricks are undoubtedly as old as playing cards themselves. Although cards were used for gaming purposes in ancient Egypt, it was not until the fourteenth century that decks could be made from linen paper, and not until the early fifteenth century that card-playing became widespread in Europe. Tricks with cards were not recorded until the seventeenth century and books dealing entirely with card magic did not appear until the nineteenth. As far as I am aware, no book has yet been written dealing exclusively with card tricks based on mathematical principles.

The earliest discussion of card magic by a mathematician seems to be in PROBLÈMES PLAISANS ET DÉLECTABLES, by Claud Gaspard Bachet, a recreational work published in France in

1612. Since then, references to card tricks have appeared in many books dealing with mathematical recreations.

The Curiosities of Peirce

The first, perhaps the only, philosopher of eminence to consider a subject matter so trivial as card magic, was the American logician and father of pragmatism, Charles Peirce. In one of his papers (see THE COLLECTED PAPERS OF CHARLES SANDERS PEIRCE, 1931, Vol. 4, p. 473f) he confesses that in 1860 he "cooked up" a number of unusual card effects based on what he calls "cyclic arithmetic." Two of these tricks he describes in detail under the titles of "First Curiosity" and "Second Curiosity." To a modern magician, the tricks are curiosities in a sense unintended by Peirce.

The "First Curiosity," based on one of Fermat's theorems, requires thirteen pages for the mere description of how to perform it and fifty-two additional pages to explain why it works! Although Peirce writes that he performed this trick "with the uniform result of interesting and surprising all the company," the climax is so weak in proportion to the complexity of preparation that it is difficult to believe that Peirce's audiences were not half-asleep before the trick terminated.

About the turn of the present century card magic experienced an unprecedented growth. Most of it was concerned with the invention of "sleights" (ways of manipulating cards secretly), but the development also saw the appearance of hundreds of new tricks that depended wholly or in part on mathematical principles for their operation. Since 1900 card magic has steadily improved, and at present there are innumerable mathematical tricks that are not only ingenious but also highly entertaining.

One illustration will indicate how the principle of an old trick has been transformed in such a way as to increase enormously its entertainment value. W. W. Rouse Ball, in his MATHEMATICAL RECREATIONS, 1892, describes the following effect:

Sixteen cards are placed face up on the table, in the form of a square with four cards on each side. Someone is asked to

select any card in his mind and tell the performer which of the
four *vertical* rows his card is in. The cards are now gathered
by scooping up each vertical row and placing the cards in the
left hand.   Once more the cards are dealt to the table to form
a square.   This dealing is by *horizontal* rows, so that after
the square is completed, the rows which were vertical before
are now horizontal.   The performer must remember which of
these rows contains the chosen card.

Once more the spectator is asked to state in which *vertical*
row he sees his card.   The intersection of this row with the
horizontal row known to contain the card will naturally enable the magician to point to the card instantly.   The success
of the trick depends, of course, on the spectator's inability to
follow the procedure well enough to guess the operating principle.   Unfortunately, few spectators are that dense.

The Five Poker Hands

Here is how the same principle is used in a modern card
effect:

The magician is seated at a table with four spectators.   He
deals five hands of five cards each.   Each person is asked to
pick up his hand and mentally select one card among the five.
The hands are gathered up and once more dealt around the
table to form five piles of cards.   The magician picks up any
designated pile and fans it so that the faces are toward the
spectators.   He asks if anyone sees his selected card.   If so,
the magician (without looking at the cards) immediately pulls
the chosen card from the fan.   This is repeated with each
hand until all the selected cards are discovered.   In some
hands there may be no chosen cards at all.   Other hands may
have two or more.   In all cases, however, the performer finds
the cards instantly.

The working is simple.   The hands are gathered face down,
beginning with the first spectator on the left and going around
the table, the magician's own hand going on top of the other
four.   The cards are then redealt.   Any hand may now be
picked up and fanned.   If spectator number *two* sees his
selected card, then that card will be in the *second* position

from the top of the fan. If the *fourth* spectator sees his card, it will be the *fourth* in the hand. In other words, the position of the chosen card will correspond to the number of the spectator, counting from left to right around the table. The same rule applies to each of the five hands.

A little reflection and you will see that the principle of intersecting sets is involved in this version exactly as in the older form. But the newer handling serves better to conceal the method and also adds considerably to the dramatic effect. The working is so simple that the trick can be performed even while blindfolded, a method of presentation that elevates the trick to the rank of first-class parlor magic.

In the pages to follow we shall consider representative samples of modern mathematical card tricks. The field is much too vast to permit an exhaustive survey, so I have chosen only the more unusual and entertaining effects, with a view to illustrating the wide variety of mathematical principles that are employed. Although most of these tricks are known to card magicians, very few of them have found their way into the literature of mathematical recreations.

TRICKS USING CARDS AS COUNTING UNITS

Under this heading we shall consider only the type of trick in which cards are used as units, without combination with other properties of the deck. Any collection of small objects, such as coins, pebbles, or matches could be similarly employed, but the compactness of cards makes them easier than most objects to handle and count.

The Piano Trick

The magician asks someone to place his hands palm down on the table. A pair of cards is inserted between each two adjacent fingers (including the thumbs), with the exception of the fourth and fifth fingers of the left hand. Between these fingers the magician places a *single* card.

The first pair on the magician's left is removed, the cards separated, and placed side by side on the table. The next pair is treated likewise, the cards going on top of the first two. This is continued with all the pairs, forming two piles of cards on the table.

The magician picks up the remaining single card and asks, "To which pile shall I add this *odd* card?" We will assume that the pile on the left is designated. The card is dropped on this pile.

The performer announces that he will cause this single card to travel magically from the left pile to the pile on the right. The left pile is picked up and the cards dealt off by pairs. The pairs come out even, with no card left over. The right hand pile is picked up and the cards taken off by pairs as before. After all the pairs are dealt, a single card remains!

Method: The working is due to the fact that there are seven pairs of cards. When these pairs are separated, each pile will contain seven cards—an *odd* number. Adding the extra card, therefore, makes this an *even* pile. If the cards are dealt by pairs, without counting them aloud, no one will notice that one pile contains one more pair than the other.

This trick is at least fifty years old. It is known as "The Piano Trick" because the position of the spectator's hands suggests playing a piano.

The Estimated Cut

The performer asks someone to cut a small packet of cards from the deck. He then cuts a larger packet for himself. The magician counts his cards. We will assume he has twenty. He now announces, "I have as many cards as you have, plus four cards, and enough left to make 16." The spectator counts his cards. Let us say he has eleven. The magician deals his cards to the table, counting up to eleven. He then places four cards aside, according to his statement, and continues dealing, counting 12, 13, 14, 15, 16. The sixteenth card is the last one, as he predicted.

The trick is repeated over and over, though the prediction varies each time in the number of cards to be placed aside—

sometimes three, sometimes five, and so on. It seems impossible for the magician to make his prediction without knowing the number of cards taken by the spectator.

Method: It is not necessary for the performer to know the number held by the spectator. He simply makes sure that he takes more cards than the other person. He counts his cards. In the example given, he has twenty. He then arbitrarily picks a small number such as 4, subtracting it from 20 to get 16. The statement is worded, "I have as many cards as you have, plus four extra cards, and enough left to make sixteen." The cards are counted, as previously explained, and the statement proves to be correct.

The method of counting seems to involve the spectator's number, though actually the magician is simply counting his own cards, with the exception of the four which are placed aside. Varying the number to be set aside each time serves to impress the spectator with the fact that somehow the formula is dependent on the number of cards he is holding.

TRICKS USING THE NUMERICAL VALUES

Findley's Four-Card Trick

A deck of cards is shuffled by the audience. The magician places it in his pocket and asks someone to call out any card that comes to mind. For example, the Queen of Spades is named. He reaches into his pocket and removes a spade. This, he explains, indicates the suit of the chosen card. He then removes a four and an eight which together total 12, the numerical value of the queen.

Method: Previous to showing the trick, the magician removes from the deck the Ace of Clubs, Two of Hearts, Four of Spades, and Eight of Diamonds. He places these four cards in his pocket, remembering their order. The shuffled pack is later placed in the pocket beneath these four cards so that they become the top cards of the pack. The audience, of course, is not aware of the fact that four cards are in the magician's pocket while the deck is being shuffled.

Because the four cards are in a doubling series, each value

twice the previous one, it is possible to combine them in
various ways to produce any desired total from 1 to 15. More-
over, each suit is represented by a card.

The card with the appropriate suit is removed from the
pocket first. If this card is also involved in the combination
of cards used to give the desired total, then the additional card
or cards are removed and the values of all the cards are added.
Otherwise, the first card is tossed aside and the card or cards
totaling the desired number are then taken from the pocket.
As we shall see in later chapters, the doubling principle involved in this trick is one that is used in many other mathematical magic effects.

Occasionally one of the four cards will be named. In this
case the magician takes the card itself from his pocket—a
seeming miracle! The trick is the invention of Arthur Find-
ley, New York City.

A Baffling Prediction

A spectator shuffles the deck and places it on the table. The
magician writes the name of a card on a piece of paper and
places it face down without letting anyone see what has been
written.

Twelve cards are now dealt to the table, face down. The
spectator is asked to touch any four. The touched cards are
turned face up. The remaining cards are gathered and returned to the *bottom* of the pack.

We will assume the four face-up cards to be a three, six, ten,
and king. The magician states that he will deal cards on top
of each of the four, dealing enough cards to bring the total of
each pile up to ten. For example, he deals seven cards on the
three, counting "4, 5, 6, 7, 8, 9, 10." Four cards are dealt on
the six. No cards are dealt on the ten. Each court card
counts as ten, so no cards are placed on the king.

The values of the four cards are now added: 3, 6, 10, and
10 equals 29. The spectator is handed the pack and asked to
count to the 29th card. This card is turned over. The magician's prediction is now read. It is, of course, the name of the
chosen card.

Method: After the deck is shuffled the magician casually notes the *bottom* card of the pack. It is the name of this card that he writes as his prediction. The rest works automatically. Gathering the eight cards and placing them on the bottom of the pack places the glimpsed card at the 40th position. After the cards are properly dealt, and the four face-up cards totaled, the count will invariably fall on this card. The fact that the deck is shuffled at the outset makes the trick particularly baffling.

It is interesting to note that in this trick, as well as in others based on the same principle, you may permit the spectator to assign any value, from 1 to 10, to the jacks, kings, and queens. For example, he may decide to call each jack a 3, each queen a 7, and each king a 4. This has no effect whatever on the working of the trick, but it serves to make it more mysterious. Actually, the trick requires only that the deck consist of 52 cards—it matters not in the least what these cards are. If they were all deuces the trick would work just as well. This means that a spectator can arbitrarily assign a new value to any card he wishes without affecting the success of the trick!

Further mystification may be added by stealing two cards from the pack before showing the trick. In this case ten cards are dealt on the table instead of twelve. After the trick is over, the two cards are secretly returned to the pack. Now if a spectator tries to repeat the trick exactly as he saw it, it will not work.

Henry Christ's Improvement

A few years ago Henry Christ, New York City amateur magician, made a sensational improvement on this effect. As in the original version, the count ends on the card that is ninth from the bottom of the deck. Instead of predicting this card, however, the spectator is allowed to select a card which is then brought to the desired position in the following manner. After the pack has been shuffled, the magician deals nine cards in a face-down heap on the table. A spectator selects one of these cards, notes its face, then returns it to the top of the heap. The deck is replaced on the pile, thus bringing the chosen card to the position of ninth from the bottom.

The spectator now takes the pack and starts dealing the cards one at a time into a face-up pile, at the same time counting aloud and *backward* from 10 to 1. If he by chance deals a card that corresponds with the number he is calling (for example, a four when he counts 4), then he stops dealing on that pile and starts a new pile next to it. If there is no coincidence of card and number by the time he reaches the count of 1, the pile is "killed" by covering it with a face-down card taken from the top of the deck.

Four piles are formed in this manner. The exposed top cards of the piles which have not been "killed" are now added. When the spectator counts to that number in the deck, he will end the count on his selected card. This is a much more effective handling than the older version because the selection of the cards to be added seems to be completely random and the compensation principle involved is more deeply hidden. The trick was first described in print by John Scarne as trick No. 30 in his book SCARNE ON CARD TRICKS, 1950. (See also Trick No. 63 for a slightly different handling by the Chicago magician Bert Allerton.)

The Cyclic Number

Many number curiosities can be presented effectively as card tricks. Consider for instance the following trick published in 1942 by magician Lloyd Jones of Oakland, California. It is based on the "cyclic number" 142857. If this number is multiplied by any number from 2 through 6, the result will contain the same digits in the same cyclic order.

The effect is as follows. The spectator is handed five red cards bearing the values of 2, 3, 4, 5, and 6. The magician holds six black cards arranged so their values correspond to the digits in the number 142857. Both magician and spectator shuffle their respective packets. Actually, the magician "false shuffles" his six cards, keeping them in the original order. (An easy way to do this is to overhand shuffle the packet twice, drawing off the cards one by one with the left thumb. Done rapidly, this gives the impression of a shuffle, though all it does is reverse the order of cards twice, thus leaving them as before.)

The magician deals his cards face up in a row on the table, forming the number 142857. The spectator now draws at random one of his five cards and places it face up beneath the row. Using pencil and paper he multiplies the large number by the value of the card he selected. While he is doing this the magician assembles the six black cards, cuts them once, and leaves them on the table in a face-down pile. After the result of the multiplication is announced, the magician picks up the pile of black cards and once more deals them in a face-up row. They form a six-figure number which corresponds exactly to the result obtained by the spectator.

Method: The black cards are picked up in their original order. It is now a simple matter for the magician to determine the spot at which these cards must be cut. For example, if the spectator is multiplying the original number by 6, the result must end with 2 because 6 times 7 (the last digit in the cyclic number) is 42. So he merely cuts the packet to bring a two to the bottom. When the cards are later dealt to form a row, the two will be the last card dealt and the number will be the same as the spectator's answer. (For Dr. E. G. Ervin's earlier version in which the cyclic number is written down and the multiplier obtained by rolling a die, see ANNEMANN'S PRACTICAL MENTAL EFFECTS, 1944, p. 106.)

The cyclic number 142857, incidentally, is the reciprocal of the prime number 7. That is, it is obtained by dividing 1 by 7. If this is done, the cyclic number appears as an endlessly repeating decimal series. Larger cyclic numbers can be found in a similar manner by dividing 1 by certain higher primes.

The Missing Card

While the magician's back is turned, someone takes a card from the deck, puts the card in his pocket, then shuffles the deck. The magician now turns around, takes the pack and deals the cards one at a time into a face-up pile on the table. After all the cards are dealt, he names the card that is missing.

Method: The value of the missing card can be determined by keeping a running total of the values of each card as they are dealt. Jacks are 11, queens 12. Kings are considered zero and ignored completely. Without the kings, the total of all card values is 312. Hence to obtain the value of the missing card, subtract the total of the 51 cards from 312. If the total is 312, then the missing card must be a king.

In adding the values, remember that to add 11 you merely add 10 and then one more. Similarly, to add 12 you add 10 and two more. Additional speed may be gained by "casting out twenties" as you go along. In other words, as soon as the total passes 20, drop 20 and recall only the remainder. After the last card is dealt you should have in mind a number from 0 to 12 inclusive. Subtract this from 12 to obtain the value of the missing card. If the deal ends with a total of 12, then the missing card is a king. (Casting out twenties is, for me, the easiest way to handle this, but many performers prefer to cast out thirteens. Thus if you add 8 and 7, drop 13 from the total and remember 2. Instead of adding 11 for a jack and then dropping 13, it is simpler to add nothing and drop 2. For a queen, drop 1. Kings are of course ignored. At the finish, subtract from 13 to find the value of the missing card.)

Once you learn the value of the card you can, of course, deal through a second time to learn its suit. But this makes the working of the trick obvious. How, then, can the suit be determined by the same deal which determines the value?

One method, which is difficult unless you are skillful in adding rapidly in your head, is to keep in mind a second running total for the suits. Spades may be given a value of 1, clubs 2, hearts 3. Diamonds are considered zero and therefore ignored. As you add, cast out tens so that at the finish you will have in mind a figure from 5 to 8 inclusive. Subtract this figure from 8 to obtain the suit of the missing card.

Jordan's Method

Another method of keeping track of the totals of both values and suits was suggested by the American magician Charles T. Jordan. Agree upon an order of suits —say, spades, hearts, clubs, diamonds. Before dealing the first card, repeat to yourself 0–0–0–0. If the first card is the Seven of Hearts, start reciting, over and over, 0–7–0–0. If the next card is the Five of

Diamonds, the recitation changes to 0–7–0–5. In other words, keep a running total of all four suits in mind. If one card only has been removed, the kings must be included in all four running totals. The final total for each suit should be 91, but because of the missing card, one total will be less. Thus if you finish 91–91–90–91, you know the missing card is the Ace of Clubs. As before, you may speed the addition by casting out twenties. When this is done, the final total is subtracted from 11 to obtain the missing card, unless the total is more than 11, in which case it is taken from 31 (or perhaps it is easier simply to remember that final totals of 20, 19, and 18 indicate a jack, queen, and king respectively).

An advantage of the Jordan method is that you can have four cards removed, one of each suit, and name all four of them as easily as you can name one. In this version, the kings may be omitted because you know that one card must be missing from each suit. The final total for each suit (ignoring kings) should be 78. If you cast out twenties, it will be 18. Hence a final result of 7–16–13–18 tells you the four missing cards are the Jack of Spades, Two of Hearts, Five of Clubs, and King of Diamonds.

It is not easy, however, to keep four or even two running totals in mind. To eliminate this complexity, I once devised a simple method of using the feet as a secret counting device. If you are seated at a table as you deal, your feet are usually hidden from view and there is little likelihood that the slight movements which are necessary will be observed.

At the start of the deal, place both feet flat on the floor. As each card is dealt, raise or lower the toes of your shoes according to the following system.

If the card is a spade, raise or lower the toe of your left foot. That is, raise it for the first spade, lower it for the second spade, raise it for the third, and so on.

If the card is a heart, raise or lower the toe of your right foot.

If the card is a club, alter the positions of both feet simultaneously.

If the card is a diamond, ignore it as far as movements of the feet are concerned.

After the last card is dealt, you determine the suit of the missing card as follows.

If your left foot is flat on the floor, the card is red.   If the left toe is raised, the card is black.

If your right foot is flat on the floor, the card is either a spade or diamond.   If the right toe is raised, the card is either a club or heart.

With this information, the suit is readily identified.   If both feet are flat on the floor, you know the missing card is a diamond.   If both toes are in the air, the card is a club.   If the left toe only is raised, it is a spade, and if only the right toe is in the air, it is a heart.

In *Hugard's Magic Monthly*, November, 1948, I proposed the use of the fingers as counting devices for the values of the cards.   In this case the dealing must be done slowly by a spectator while you rest a hand on each thigh.   The fingers are numbered left to right from 1 to 10.   As each card is dealt, raise or lower the appropriate finger.   Jacks are kept track of by sliding the left hand forward or back along the leg.   Queens by sliding the right hand.   Kings are ignored.   The suits may be followed by using the feet as previously explained.

The finger method permits you to determine the values of several cards removed from the deck provided no two cards have the same value.   You merely note which fingers are raised and/or which hand is forward on the leg at the end of the deal.   You must of course know how many cards have been removed, because the only way to determine a king is by having one card unaccounted for.   Other card effects can be based on the finger-counting, as I have suggested in the article previously cited.

TRICKS BASED ON DIVISION OF COLORS AND SUITS

Stewart James' Color Prediction

The magician writes a prediction on a piece of paper and places it aside.   After a spectator has shuffled the deck he is asked to deal the cards to the table by taking them off in pairs, turning each pair face up.   If both cards are black, he is told to place them in a pile on the right.   If both cards are red, they are placed to the left.   If they are mixed—that is, one red and one black—they are tossed into a discard pile.   This continues until the entire pack has been dealt.   The magician stresses the fact that pure chance determines the number of cards in the red and black piles.

After all cards are dealt, the red and black piles are counted. The prediction is now read.   It states, "There will be four more red cards than black."   This is correct.

The cards are gathered, shuffled, and the trick repeated. The prediction this time reads, "There will be two more black cards than red."   This also proves correct.

On the third and last repetition, the red and black piles have an equal number of cards.   The prediction reads, "The two piles will be exactly the same."

Method: Before beginning the trick the magician secretly removes four black cards from the deck.   If he is seated at a table the cards can be held in his lap.

After the first deal by pairs is completed there always will be four more cards in the red pile than in the black.   The reason for this is, of course, that the discarded pairs are half red and half black, thereby removing an equal number of reds and blacks from the deck.   Since the deck is short by four black cards, the red pile must of necessity have four more cards in it than the black.

While attention is focused on the counting of the two piles, the magician casually picks up the discard pile of mixed reds and blacks, and holds it in his lap.   He secretly replaces the four black cards he has been holding on his lap, and steals two red cards.   When all the cards are later gathered and shuffled, the deck will be short two red cards in preparation for the second prediction.

The same procedure is followed in setting the deck for the third and final prediction.   This time the two red cards are replaced but no cards are stolen.   The deck now contains its full 52 cards, therefore the red and black piles will be equal, and if anyone should count the cards after the trick is over, he will find a complete pack.   Stewart James, a magician in

Courtright, Ontario, Canada, contributed this fine effect to *The Jinx*, September, 1936.

The Royal Pairs

The magician removes the kings and queens from the deck. The kings are placed in one pile, the queens in another. The piles are turned face down and one placed on top of the other. A spectator cuts the packet of eight cards as often as he wishes. The magician holds the packet behind his back. In a moment he brings forward a pair of cards and tosses them face up on the table. They prove to be a king and queen of the same suit. This is repeated with the other three pairs.

Method: When the two piles are formed, the magician makes sure that the order of suits is the same in each pile. Cutting will not disturb this rotation of suits. Behind his back he simply divides the packet in half, then obtains each pair by taking the top card of each half. These two cards will always be a king and queen of the same suit.

TRICKS USING FRONT AND BACK

Matching the Colors

The deck is cut into two halves. One half is turned face up and shuffled into the other half, which remains face down. The deck now consists of a mixture of face-up and face-down cards. This mixed pack is thoroughly shuffled by a spectator.

The magician extends his right palm and the spectator is asked to deal 26 cards onto the palm. After this is done the performer states that the number of face-up cards in his hand is exactly equal to the number of face-up cards in the packet retained by the spectator. Both halves of the deck are spread on the table and the face-up cards in each half are counted. The statement proves correct. Although the probability is high that the face-up cards in each half will be *approximately* the same in number, it is extremely improbable that they will be *exactly* the same. Yet the trick may be repeated any number of times and this is always the case.

Method: Before starting the trick, the magician secretly notes the 26th card in the deck. This enables him to cut the deck *exactly* in half. He has only to fan the cards, faces showing, and divide them at the card he is remembering. To the audience it looks as if he merely cut the deck carelessly into two approximately equal groups.

One half is turned face up and the two halves are shuffled together. The spectator now counts 26 cards onto the magician's palm. A little reflection and you will see that the 26 cards on the magician's hand must contain a number of *face-down* cards exactly equal to the number of *face-up* cards in the remaining half. It is only necessary, therefore, for the magician surreptitiously to turn his packet over. When he turns his hand palm down to spread the cards on the table, such a reversal occurs automatically. This should be done while the spectator is occupied with spreading his own half, making him less likely to notice that the magician has turned his cards over in the process of spreading them. Because of the reversal, the number of face-up cards in each half will be exactly the same.

Before repeating the trick, the magician must remember to reverse one of the halves (it does not matter which) again. This restores the deck to its former condition—26 face-up cards and 26 face-down.

This fine effect is the invention of Bob Hummer, a magician now living near Perryman, Maryland, and may be found in one of his many booklets of original tricks. Hummer applies the same principle to a trick using ten cards only, five red and five black. The cards of one color are reversed and all ten cards shuffled thoroughly by a spectator. The magician takes the packet momentarily behind his back. He then brings the cards forward, five in one hand and five in the other, and spreads each group on the table. The number of face-up cards in each group is the same, and they are of opposite color. For example, if there are three face-up red cards in one group of five, there will be three face-up black cards in the other group. The trick may be repeated as often as desired, always with the same result.

The method of working is essentially the same as before.

Behind his back the magician simply divides the cards in half, then reverses either half before he brings the two groups forward. Any even number of cards may be used of course, provided that half are red and half black. The trick was marketed by Harold Sterling, Royal Oak, Michigan, under the title of "The Gremlins," with ten cards, five bearing pictures of red gremlins and five bearing pictures of green gremlins.

Hummer's Reversal Mystery

The magician hands a packet of eighteen cards to a spectator with the request that he hold them under the table, out of sight, and mix the cards by performing the following operations. He is to turn the top *pair* of cards face up, then cut the packet. Again, he is to reverse the top pair and follow with a cut. This is repeated as often as he wishes. Obviously, this process will mix the cards in an unpredictable fashion, causing an unknown number of cards to be face up at various positions in the packet.

The magician is seated at the opposite side of the table from the spectator. He reaches under the table and takes the packet. Keeping his hands beneath the table, so that the cards are hidden from everyone, including himself, he states that he will name the number that are face up in the pack. He announces a figure. When the cards are brought into view and spread on the table, the figure proves to be correct.

A second effect now follows. The magician arranges the eighteen cards in a special way, without letting the spectator see the arrangement. The packet is handed to him with the request that he hold it beneath the table and destroy the arrangement by repeating the same mixing operation that he employed before.

After the spectator has turned pairs and cut the packet enough times to feel that the cards have been thoroughly mixed, the performer stands up and turns his back to the table. He asks the spectator to bring the packet into view and note the top card. If this card is face down, he is to turn it face up. If the card is already face up, he is to turn it face down. In either case he must remember the card. The packet is then given a single cut.

After this is done the magician seats himself and again reaches under the table to take the packet. He states that he will attempt to find the chosen card. A moment later he brings the cards above the table and announces that he has straightened out the packet so that all the cards are now face down except for one card—the selected one. The spectator names his card. The performer spreads the packet on the table. All the cards are face down except for the chosen card which is face up in the center of the spread!

Method: The operation of this startling trick is entirely mechanical. For the first effect, the magician has only to take the packet beneath the table and run through it reversing every other card. He then states that the packet contains *nine* face-up cards, naming a figure that is half the total. (Any even number of cards can be used for this trick.) This will turn out to be correct.

In preparing the packet for the second effect, the performer arranges it so that every other card is face up. Of course he does not let the audience see that this is the arrangement. After the card has been selected in the manner described, the magician takes the packet beneath the table and follows exactly the same procedure as before—that is, reversing every other card. This will cause all the cards to face one way except for the chosen card, which will be reversed near the center.

This trick may also be found in Bob Hummer's FACE-UP FACE-DOWN MYSTERIES. It has been subjected to many variations and additions. For the second part of the effect Eddie Marlo, a Chicago amateur conjuror, suggests having a prepared packet of eighteen cards, arranged with alternate face-up and face-down cards, concealed on your lap or on the chair beneath your thigh. The original packet is handed to the spectator under the table, and in the act of doing so, it is switched for the prepared one.

Oscar Weigle, New York City amateur magician, proposed beginning the trick by removing eighteen cards from the deck apparently at random but actually taking alternate red and black cards. The packet is given a series of false overhand shuffles, each time running an odd number of cards before throwing the remainder on top.   This shuffles the packet without disturbing the alternate red-black order.   The first part of the trick is now presented, but after showing the nine face-up cards you call attention to the fact that the spectator has also miraculously separated the reds from blacks.   All the face-up cards are one color and the face-down cards of the opposite color.   I explained Weigle's variation in more detail in *Hugard's Magic Monthly*, November, 1948.   The following year Mr. Weigle published his pamphlet COLOR SCHEME, a further elaboration of the effect.

The Little Moonies

Hummer's pamphlet, mentioned above, contains another unusual application of the principle—a set of cards which he calls "The Little Moonies."   Each card pictures a smiling face, but when turned upside-down the face appears to be frowning.   The cards (there must be an even number) are arranged so that they are alternately smiling and frowning, although the spectator does not know this.   The packet is cut as often as desired.   The spectator now marks with pencil the back of the top card.   This card is placed second from top and the card now on top is marked in a similar manner.   The cards are cut again then handed under the table to the performer.   A moment later he spreads the cards on the table.   All the faces will be smiling except two, or all the faces will be frowning except two.   These two cards are turned over and they are seen to be the two which were marked.

The working of the trick is as follows: Under the table the cards are "dealt" into two piles, using the crotch of left thumb and forefinger to hold one pile and the crotch of first and second fingers to hold the other pile.   Either pile is then reversed end for end.   This automatically gives the same facial expression to all the cards except the two marked ones.

The next two chapters will be devoted to tricks that illustrate the last of the five card properties mentioned at the outset, namely the ease with which cards can be given orderly arrangements.

Chapter Two

Tricks with Cards—Part Two

Although many of the tricks described in the previous chapter clearly involve an "ordering" of cards, it seemed more convenient to classify them under other heads. In this chapter we shall consider tricks in which the arrangement of cards plays a major role. In most cases, other properties of the deck also come into play.

O'Connor's Four-Ace Trick

The magician asks someone to name a number between 10 and 20. He deals that number of cards into a pile on the table. The two digits in the number are now added and the corresponding number of cards is taken from the top of the pile and placed back on top of the deck. The top card of the pile is now placed aside, face down. The pile is returned to the top of the deck. Another number between 10 and 20 is requested and the process just described is repeated. This continues until four cards have been selected in this curious manner. The four cards are now turned face up. All four are aces!

Method: Before the trick begins the aces must be in the ninth, tenth, eleventh, and twelfth positions from the top of the deck. The trick then works automatically. (Billy O'Connor contributed this effect to the *Magic Wand*, June–Sept., 1933.)

The Magic of Manhattan

A spectator is asked to cut the deck near the center without completing the cut, then pick up either half. He counts the cards in this half. Let as assume there are 24. The 2 and 4 are added to make 6. He looks at the sixth card from the *bottom* of the half-deck he is holding, then replaces the half-deck on the other half, squares the pack, and hands it to the magician. The magician starts dealing the cards from the top, spelling aloud the phrase "T-H-E M-A-G-I-C O-F M-A-N-H-A-T-T-A-N," one letter for each card dealt. The spelling terminates on the selected card.

Method: The described procedure always places the chosen card nineteenth from the top of the pack. Therefore any phrase of nineteen letters will spell to the chosen card. Bill Nord, the New York City amateur conjuror who invented this effect, suggested "The Magic of Manhattan," but any phrase of nineteen letters will of course work just as well.

Both this trick and the preceding one are based on the fact that if you add the digits in a number and subtract the total from the original number, the result will always be a multiple of nine.

Predicting the Shift

A packet of thirteen cards is cut several times, then given to a spectator. The magician turns his back and requests that the spectator transfer, one at a time, any number of cards from 1 to 13, inclusive, from the bottom to the top of the packet.

The magician turns around, takes the packet, fans it face down, and immediately draws out a card. When the card is turned over, its face value will correspond to the number of cards shifted. The trick may be repeated indefinitely.

Method: The packet contains one card of each value from ace to king, arranged in numerical progression with the king on top. The packet is cut several times, but as the performer hands it to the spectator, he notes the bottom card. Assume it is a four. After the cards have been shifted, the magician counts to the *fourth* card from the top, and turns it face up. The value of this card will correspond to the number shifted.

The trick is repeated by noting the bottom card again as the packet is handed out. Better still, by knowing the order of rotation (which remains the same regardless of cutting and shifting), the magician simply counts backward from the card he turned over, to the bottom card. In this way, he learns the bottom card without having to glimpse it.

The Keystone Card Discovery

The deck is shuffled. The magician glances through it for a moment, then places it face down on the table and names a card. For example, he names the Two of Hearts. Someone now calls out a number between 1 and 26. The performer counts that number of cards one at a time to the table, then turns over the card on top of the pack. But this card is *not* the Two of Hearts.

The magician looks puzzled. He states that perhaps the card is in the lower half of the deck. The incorrect card is turned face down on the pack, and the cards on the table are replaced on top. The spectator is asked to name another number, this time between 26 and 52. Once more this number of cards is counted to the table. Again the card on top of the deck is *not* the Two of Hearts.

The incorrect card is turned face down on the deck and the cards on the table are returned to the top of the pack. The magician now suggests that perhaps the Two of Hearts can be found by subtracting the first number from the second. This is done. A number of cards equal to the remainder is now dealt. When the card on top of the deck is turned over, it is the Two of Hearts!

Method: After the magician looks through the deck he simply names the top card of the pack. Counting the cards twice to the table will automatically place this card in the position indicated by the difference between the two designated numbers.

The trick was sold in 1920 by Charles T. Jordan as "The Keystone Card Discovery," and apparently re-invented by T. Page Wright, who described it as an original effect in *The Sphinx*, December, 1925.

Two-Pile Location

The magician turns his back with the request that someone deal the cards into two small piles, an equal number of cards in each.   When this is done, the spectator is told to look at the card now on top of the deck.   One pile is replaced on the deck, on top of the selected card.   The other pile is placed in the spectator's pocket.

The magician turns around and takes the pack, placing it momentarily behind his back.   The cards are then brought forward and placed on the table.   The spectator removes the packet from his pocket and adds them to the top of the deck. The magician points out that he had no way of knowing the number of cards in the spectator's pocket, therefore any arrangement of the deck made behind his back would seem to be modified by this addition of an unknown quantity of cards.

The spectator is now asked to pick up the deck and deal one card at a time from the top.   As the cards are dealt, he spells aloud the phrase, "T-H-I-S I-S T-H-E C-A-R-D I S-E-L-E-C-T-E-D," calling a letter for each card dealt.   The spelling terminates on a card which proves to be the chosen one!

Method: Behind his back the performer counts off cards from the top of the deck into his right hand, spelling whatever sentence or phrase he wishes to use later.   This reverses the cards counted.   They are replaced back on the deck. Later, when the spectator replaces the cards he has kept in his pocket, the chosen card will automatically be brought to the correct position for the final spelling.

The phrase used for the spelling must contain a number of letters greater than the number of cards in each pile.   For this reason the spectator should be told not to exceed a certain number when he forms the two piles.   An effective variation of the trick is to use the spectator's own name for the spelling.

Spelling the Spades

The magician asks the spectator to give the deck a single riffle shuffle, followed by a cut.   The performer takes the deck and runs through it face up, removing all the spades.   This packet of spades is handed to the spectator, face down, with the request that he shift one card at a time from top to bottom, spelling A C-E.   The card on the last letter is turned face up.

It is the Ace of Spades. The ace is tossed aside and the same spelling procedure repeated for T-W-O. Again, the Two of Spades turns up at the end of the spelling. This continues until all the spades from ace to king have been spelled in this manner.

Method: The magician prepares for the trick in advance by arranging the thirteen spades as follows. He holds a king in his left hand. He then picks up the queen and places it on the king, saying "Q" to himself. The cards are now shifted one at a time from *bottom* to *top* as he spells "U-E-E-N." The same procedure is followed for jack, ten, nine, and so on down to ace. In other words, he simply performs in reverse the spelling process which the spectator will perform later. At the finish, he will have a packet of thirteen cards so arranged that they will spell from ace to king in the manner already described.

This packet is placed in the center of the deck and the trick is ready to begin. The spectator's riffle shuffle will not disturb the order of the series. It will merely distribute an upper portion of the series into the lower part of the deck and a lower portion of the series into the upper part of the deck. A cut made near the center of the pack will now bring the series back to its original order, although the cards will of course be distributed throughout the deck. As the magician runs through the pack to remove the spades, he takes the cards one at a time, beginning at the bottom of the deck. As each spade is removed it is placed face down on the table to form a packet of thirteen cards. This packet will then be in proper order for the spelling. The trick was sold in 1920 by Charles T. Jordan, under the title of "The Improved Chevalier Card Trick."

Because it is such a simple matter to arrange cards for spelling tricks of this type, scores of such effects have been published. An amusing variation is one in which the magician spells the cards correctly but at intervals hands the packet to a spectator with the request that he spell the next card. Invariably the spectator gets the wrong card, or perhaps the Joker, although the card appears at the proper spot when the magician takes back the packet and spells the name of the card again. (For an early example, see *The Jinx Winter Extra,*

1937–38, p. 273.) A similar variation is for the magician occasionally to spell a card incorrectly. When this is done, the correct card does not turn up. But when the magician corrects the spelling and tries again, he finds the card. (See Trick No. 65 in SCARNE ON CARD TRICKS.)

Elmsley's Card Coincidence

A deck is shuffled and cut into two approximately equal halves. A spectator looks at the top card of either half, replaces it, and cuts the pile. Another spectator does the same thing with the other half. Each spectator remembers the card he noted. The magician now looks briefly through each pile, cutting it before he replaces it on the table.

The two piles are now side by side on the table, each containing a selected card. Using both hands, the magician starts taking cards simultaneously from the tops of both piles and dealing them to the table beside each pile. One hand, however, deals the cards face up while the other hand deals them face down. He asks to be told as soon as one of the selected cards is turned face up. When this occurs, he pauses, holding the face-up selected card in one hand and a face-down card in the other. The second chosen card is named. When the face-down card is turned over, it proves to be the second selected card.

Method: After the deck is shuffled, look through the cards quickly and memorize the bottom card and the card which is 27th from the bottom. Place the pack face down on the table and let the spectator cut it as many times as he wishes to make sure that you do not know the top or bottom cards of the deck. The deck is now cut into two approximately equal piles. If both "key" cards should happen to fall in one pile, the trick will not work, but since the cards are 26 cards apart, the odds are high against this happening.

Spectators now look at the top card of each pile, giving the pile a cut after the card is returned. Because of the random cuts, it seems impossible that you would have any clues as to the positions of the selected cards.

Pick up either pile. As you look through it, count the number of cards and remember the number. Then find one of your key cards and cut the pile so as to bring this card to the top. Replace the pile face down.

Pick up the other pile and look for the other key card. If the first pile happened to have exactly 26 cards in it, merely cut the second so the key card goes on top. If, however, the first pile contained more or less than 26, the cut must be made a trifle above or below the key. Since the procedure is slightly different in the two cases, we will illustrate each by an example.

If the first pile contains *less* than 26 cards, subtract its number from 26. The difference tells you the position to which the key card must be brought from the *bottom* of the second pile. For example, suppose the first pile contains 22 cards. This is 4 less than 26. Cut the second pile to bring the key card to the fourth position from the bottom.

If the first pile contains *more* than 26, subtract 26 from the number. The answer, *plus 1*, tells you the position to which the key card must be brought from the *top* of the second pile. For example, suppose the first pile contains 28 cards. This is 2 more than 26. Adding 1 makes it 3. Cut the second pile to bring the key card to the third position from the top.

After you have cut each pile at the proper spot, start dealing from the top of each pile, using both hands in unison. One hand (it does not matter which) deals the cards face up, the other deals face down. Ask to be stopped as soon as one of the chosen cards appears in the hand that is dealing face up. When this occurs, pause, request the name of the other card, then slowly turn over the face-down card in the other hand. It should be the card that has just been named.

This trick is one of a number of recent effects that make use of the "center card" (either the 26th or 27th card from the top of the deck). It is the invention of Alex Elmsley, a young engineer now living in London, and was published in the British magic journal *Pentagram*, Feb., 1953. The handling given here, which differs slightly from Elmsley's, is that of the American magician Dai Vernon.

Magic by Mail

The magician sends a deck through the mail to a friend, with the following instructions.   He is to cut the deck as often as he wishes, give it *one* riffle shuffle, then cut it again as many times as desired.   He then cuts the pack into two halves, takes a card from the center of one half, notes its face, and inserts it into the center of the other half.

Either pile is now picked up, given a thorough shuffle, and mailed to the magician.   The magician is not told whether it is the half which contains the chosen card, or the half from which the card was taken.   Nevertheless, a few days later a postcard from the magician arrives, giving the name of the selected card.

Method: Before mailing the deck to his friend, the magician shuffles it, then jots down the names of the cards in the order in which they fall in the deck.   This list is regarded as a circular series, the bottom connecting with the top to form an endless chain.

When the half-deck arrives in the mail, the magician goes through the cards checking off each one on his list.   Because of the single riffle shuffle, the checked cards will fall into two interlocking "runs" of consecutive cards.   However, there will be either one card within one of the runs which is not checked, or one card will be checked which does not belong to either run.   It will be, of course, the chosen card.   The trick was invented by Charles T. Jordan, who includes it in his THIRTY CARD MYSTERIES, 1919.

Belchou's Aces

A spectator cuts the deck into four piles, which we shall call A, B, C, and D.   Pile D is the pile which formerly was the top of the deck.   He is instructed to pick up pile A and deal three face-down cards on the spot it formerly occupied on the table, then to deal one card face down upon each of the other three piles.   Pile A is then replaced on top of the three cards.

Exactly the same procedure is followed with the remaining

three piles, taking them in B, C, D order. When the top cards of the piles are turned face up, they prove to be the four aces!

Method: Unknown to the spectator, the four aces are on top of the deck at the outset. The trick will then work automatically. The trick was first described by Oscar Weigle in the *Dragon*, June, 1939, where he credits its invention to Steve Belchou, of Mount Vernon, N.Y. The working is beautifully simple, and the effect quite startling to observers.

The Tit-Tat-Toe Trick

A large tit-tat-toe board is drawn on a sheet of paper. The magician and a spectator proceed to play a game, but instead of marking on the paper, they indicate their moves by placing cards which they take from the top of a pile of nine cards. The spectator places his cards face up and the performer places his cards face down. The game ends in a draw, with all nine cards on the board. The face-down cards are now turned over to reveal a magic square. All rows—vertical, horizontal, and the two diagonals—add to fifteen.

Method: Several years ago I proposed this effect to Don Costello, New York City amateur magician and teacher of mathematics. Costello quickly realized that the trick was possible only if the magician played first, placing a five in the center of the board. His opponent would then have a choice of playing either in a corner square or on a side. In both cases the performer could force all the remaining moves. The problem was to work out a setup of nine cards that would enable the magician to force the outcome. Costello worked out several such setups, all requiring an adjustment after the spectator has indicated where he intends to make his first move.

The trick intrigued Dai Vernon who devised a subtle method by which this necessary adjustment can be made without the spectator being aware of it. What follows is Vernon's handling of the Costello trick.

Begin by arranging nine cards of one suit, say hearts, in the following order:

Ace, 8, 2, 7, 3, 4, 5, 6, 9.

The ace should be the top card of the packet. An easy way to remember the order is to arrange the cards first from ace to nine, then shift the eight and seven to the desired positions. Place the packet on the bottom of the deck and you are ready to begin.

Ask a spectator to give the deck two thorough riffle shuffles. This will not disturb the order of the nine cards. It merely distributes them through the deck. Pick up the pack and go through it face-up, stating that you will remove all the hearts from ace to nine, taking them in the order in which you find them. This procedure seems to insure a random order for the nine cards, but actually places them in the original order —ace on top and nine on the bottom when the packet is held face down.

The tit-tat-toe board is now drawn on a large sheet of paper —or you can use an imaginary board, since your first play is in the center, making it easy to visualize the positions of the other squares. You must, however, make your first play in the following manner. Fan the packet in your hands, faces toward you, and divide the fan into two parts. Your left hand holds the top six cards and your right hand the bottom three. With your right hand place the top card of its group (it will be the five) face down on the center of the board, retaining the other two cards in the hand. Keep your hands separated and ask your opponent to point to the square where he wishes to make his first play.

If he points to a corner square, replace the two cards in your right hand on *top* of the six cards in your left. If he points to a side square, replace the two cards on the *bottom*. In either case, immediately square the packet and place it face down on the table.

Ask him to remove the top card of this packet, turn it *face up*, and place it in the square he indicated. From now on each play is made by taking a card from the top of the packet and placing it on the board. Your plays are always face down, his face up. Later, he will not recall that you handled the packet at all after the first play. His recollection will be that the packet was simply placed on the table and all the moves made by taking cards from the top. This will make it extremely difficult for a mathematically-minded spectator to reconstruct the trick.

After the spectator's second play you must force all remaining moves in such manner that the cards will form a magic square. Since there are two distinct lines of play, depending on whether his first move is a corner or a side square, we will take each line of play in turn.

If he plays a corner square, make your next play on *either side* of the corner diagonally opposite the corner where he played. All remaining moves, both his and yours, will now be forced. This assumes, of course, that a player must move to prevent three in a row whenever possible. The arrows in Fig. 1 show the two squares on which you may make your second play.

If he plays a side square, then play in the corner on *either side* of his play (the two squares indicated by arrows in Fig. 2). This will force him to play in the diagonally-opposite corner. You then play in the side square which *contacts both of his previous plays*. He is now forced to play in the side square on the opposite side. Your next play is in the corner *adjacent to his last play*. The remaining moves are forced. I am indebted to Geoffrey Mott-Smith for supplying these simple rules of strategy.

For your convenience in mastering the trick, Fig. 3 shows the two possible forms of the final magic square, one a mirror-image of the other. Each square may of course be rotated to four different positions.

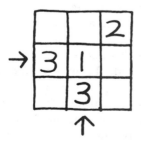

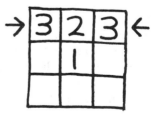

FIG. 1   FIG. 2

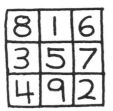

FIG. 3

Other Tricks of Interest

The card tricks explained in this chapter represent only a small fraction of the thousands of mathematical effects which have appeared in the magic literature during the past forty years. Good references for tricks of this nature are Jean Hugard's ENCYCLOPEDIA OF CARD TRICKS (the section on self-working effects); the books of Rufus Steele; the books and pamphlets of Bob Hummer; and the books and pamphlets of Stewart James. "Remembering the Future," a card trick first published by James in 1948 (currently offered for $1.00 by the Sterling Magic Company, Royal Oak, Michigan), is one of the classics of mathematical card magic. It is based on a subtle handling of "digital roots," about which something will be said in our last chapter.

John Scarne's SCARNE ON CARD TRICKS deserves special mention because of the large number of mathematical tricks which it contains. Outstanding are Stewart James' effect (No. 72) and Stewart Judah's trick (No. 112). The following tricks also impressed me as of special interest (the numbers refer to tricks, not to pages): 15, 24, 33, 34, 43, 52, 54, 67, 81, 86, 89, 94, 97, 107, 109, 115, 126, 143, 144, 150.

Scarne's book contains several mathematical effects known as "telephone tricks" because they can be performed over a telephone. Many other clever telephone tricks have appeared in print, such as Professor Cheney's effect described in Wallace Lee's MATH MIRACLES, and Richard Himber's "No Questions Asked," sold as a manuscript in 1950.

Eddie Joseph's "Staggered," which was introduced by the magic dealers in 1949, is perhaps the most influential mathematical card trick of recent years. Scores of subtle effects have been based on the principle (two decks of cards, one arranged in the reverse order of the other), or suggested by it.

The magic magazines are also gold mines of mathematical material for anyone who cares to search for it. Frank Taylor's "Double Stop" in *The Phoenix*, Feb. 24, 1950, is particularly noteworthy. Attention should also be called to the excellent series of articles which ran in *The Bat*, beginning Nov., 1948, dealing with the faro shuffle (a riffle shuffle in which the two halves interlace perfectly in alternate single cards) and its mathematical possibilities. For performers unable to execute perfect faro shuffles, the same effects in reverse can be achieved by running through a fanned deck, jogging alternate cards upward, then pulling out the jogged cards. Many tricks have been based on this reverse form of the faro, such as Dai Vernon's Trick No. 81 in Scarne's book.

An ancient and well-known mathematical trick making use of the four Latin words MUTUS, NOMEN, DEDIT, and COCIS as a mnemonic device has been greatly improved by Stewart Judah, an amateur conjuror in Cincinnati. Judah's version—using the words UNDUE, GOANO, TETRA, and RIGID—may be found in *The New Phoenix*, No. 319 (November, 1954).

Chapter Three

From Gergonne To Gargantua

The following is one of the oldest of mathematical tricks that involve the ordering of cards, and one of the most intriguing. A spectator thinks of any card in a packet of 27. The packet is held face down and dealt into three face-up piles. The spectator indicates which pile contains his card. The piles are assembled, dealt once more into three face-up groups. Again the spectator points out the pile containing his card. The procedure is repeated a third and last time, after which the magician is able to do one of three things:

1. Name the exact position of the chosen card from the top of the packet.

2. Find the chosen card at a position previously demanded by the spectator.

3. Name the card.

A special chapter is devoted to this trick because of the interest which it has aroused among mathematicians. Known as Gergonne's Pile Problem (after Joseph Diez Gergonne, the French mathematician who in 1813 was first to analyze it extensively), the trick has been much discussed in the literature of mathematical recreations. The working principles have been generalized to apply to any given number of cards (see Ball's MATHEMATICAL RECREATIONS, 1947 revised edition, p. 316). In the literature of magic the trick may be found in Professor Louis Hoffmann's MORE MAGIC, p. 32, and many earlier books on conjuring. In recent years, however, several new aspects of the trick have been developed by magicians—aspects that have not yet found their way into the literature of either conjuring or mathematical amusements.

Each of the three presentations listed above requires a different handling. We shall discuss them in turn.

Naming the Position of the Card

In this version the spectator is permitted to assemble the piles after each deal, picking them up in any order he chooses. The packet of 27 cards is always held face down for dealing and the cards are always dealt into face-up piles. The dealing may also be done by the spectator. In fact, it is not necessary for the performer to touch the cards at any time. He merely watches the procedure and after the third and final assembly, correctly states the numerical position of the chosen card in the packet of 27.

The simplest way to perform this effect is to memorize the following table:

First deal	Top	1
	Middle	2
	Bottom	3
Second deal	Top	0
	Middle	3
	Bottom	6
Third deal	Top	0
	Middle	9
	Bottom	18

Consider the face-down packet of 27 cards as consisting of three groups of nine cards each. These groups are designated on the chart as top, middle, and bottom. Each time the spectator assembles the piles, after telling you which one contains his card, note whether the designated pile falls into a top, middle, or bottom position in the *final face-down* packet. Each assembly, therefore, provides you with a key number. Adding these numbers will give you the final position of the card from the top of the packet.

For example, suppose the spectator assembles the piles after the first deal in such a way that the pile containing his card goes on the bottom. The second time he picks up the piles, the pile with his card goes in the middle, and the last time he picks them up the pile goes on top.   This gives you the key numbers 3, 3, and 0, which add to 6.   The card will therefore be sixth from the top of the packet.

The trick works by a simple process of elimination.   The first deal narrows the card to a group of nine, the second deal narrows it to three, and the last deal to one.   It is easy to see this by turning the pile containing the card face down and putting a pencil mark on the back of the chosen card so it can be followed easily.   Assemble the piles in any order and deal again.   You will observe that the nine face-down cards are now distributed throughout the three piles, each pile containing three face-down cards.   Note the group containing the marked card.   Let this group of three remain face down, but turn the other six cards face up.   Assemble and deal again. This time each of the three face-down cards is in a different pile.   Knowing the pile which contains the chosen card will therefore eliminate all possibilities except one.   The position of the chosen card after the final pick-up may be anywhere from 1 to 27, but it is easy to see that the position will be rigidly determined by the manner in which the piles are assembled after each deal.

Once you become thoroughly familiar with the selective process by which the trick operates, you may find that you do not need the chart at all to determine the final position of the card.   Simply follow the first designated pile in your mind, eliminating cards as the trick proceeds until only one card, whose position you know, remains.   With practice it is by no means difficult to perform the trick in this manner.

Bringing the Card to a Named Position

In this version the spectator is asked to state in advance the number at which he desires his card to be after the final pick-up.   The magician of course must be permitted to assemble the piles after each deal.   At the conclusion of the trick the chosen card is found at the demanded position.

The same chart may be used for this effect.   Refer to it to find three numbers, one in each dealing group, which will add to the desired number.   These three numbers tell you where to place the pile containing the chosen card as you make each pickup. More complicated charts for arriving at the same result may be found in Hoffmann's MORE MAGIC and other old books on conjuring. Performers of the last century sometimes placed these charts inside opera glasses through which they peered to obtain the necessary information!

Walker's Method

A much simpler method, not requiring a chart of any sort, was explained by Thomas Walker in the monthly organ of the Society of American Magicians, *M.U.M.*, October, 1952. Walker's method is as follows.

Let us assume the called-for position is 14 from the top. When you deal the cards for the first time, count them as you deal and note into which pile the 14th card falls. It will fall on the second pile. This tells you that when you assemble the piles for the first time, the pile containing the spectator's card must go in the second or middle position.

When you deal the second time, count again until you reach 14, but this time, instead of saying 14 to yourself, say 3. Do not count the next two cards. When you deal on the middle pile once more, say 2 to yourself. Ignore the next two cards again, counting 1 to yourself when you place the third card on the center pile. When you deal the next card on this pile, go back to 3 and repeat the series. In other words, your count is 3, 2, 1—3, 2, 1—3, 2, 1 until you finish dealing the packet. The number of the last layer of three indicates the position in which you must place the pile containing the card. In the example we are following, the number is again 2. Therefore when you assemble the piles, the pile containing the card must go in the center once more.

No counting is necessary during the third deal. Once the pile is identified you know immediately whether to place it top, middle, or bottom. If the card is to be brought to a position in the upper third of the packet, the pile naturally must go on top. If the position is in the middle third, as is 14 in our example, the pile goes in the middle. If in the lower third, it goes on the bottom.

Actually, Walker's method merely uses the dealing of the cards as a counting device for performing the calculations suggested by Professor Hoffmann. If we performed these calculations in our head, we would do so as follows:

First pickup: Divide the number called for by 3. If the remainder is 1, the pile goes in the first position. If 2, it goes in the second position. If no remainder, the third position.

Second pickup: Consider the 27 cards as divided into three groups of nine each. Each group in turn is divided into three sub-groups. Ask yourself whether the number called for is in the first, second, or third sub-group of its larger group of nine. The answer tells you whether to place the pile in first, second, or third position.

Third pickup: This is handled as previously explained.

You may have noticed that to bring the card to the 14th position, each pickup must place the designated pile in the middle, and that the 14th card is exactly in the middle of the final packet. Rules of similar simplicity hold for bringing the card to the top or bottom of the final packet. To bring the card to the top, place the pile on top in each pickup. To bring it to the bottom, place the pile each time on the bottom.

Dai Vernon has called my attention to a method of picking up the piles which, if done rapidly and in an off-hand manner, will make it appear as though you always take the piles in a left to right order. This is done by placing the right hand on each pile and sliding it toward you until it clears the edge of the table and drops into the left hand which is held at the table edge to receive the cards. You may permit the pile to drop into the left hand and remain there, or your left fingers can simply aid in squaring up the cards. In the latter case, the cards are retained by the right hand which immediately goes back to the table to pick up the next packet. These alternate procedures enable you to place the designated packet at any spot you wish. For example, let us suppose that the chosen card is in the third or right-hand pile, but you wish to bring this pile to a middle position. Pick up the first pile and slide it back to the table edge, letting it fall in the left hand. Go back for the second pile, but this time retain it in your right hand, using the left hand only for squaring up the cards.

The right hand immediately goes back to the table, slapping the second packet on top of the third, and bringing the cards back to the table edge where they drop into the left hand.

With a little experimenting you will see that the desired pile can be placed in any position you wish, although the general movements of both your hands appear to be the same in all cases. If the assembling is done rapidly, few spectators will notice how the pickup is being varied.

Naming the Card

There are many procedures that enable you to name the chosen card. One is to assemble the cards yourself in such a way as to bring the card to a desired position, then secretly note the card at that position before the final packet is assembled and placed face down. For example, assume that you plan to bring the card to third from the top. On the last deal you know that the card must be one of three cards, each of which is the third card from the top of each pile. As you deal those three cards, either memorize them or deal them carelessly so their index corners are exposed when cards are placed on top of them. As soon as the spectator points to the pile containing his card, you will know its name. In this way you can name both the card *and* its final position. If you repeat the effect, bring the card to another position, but learn its name in the same manner.

A second method requires that you begin the trick with a group of 27 cards in a known order. The spectator can be permitted to assemble the piles, because the order in which they are picked up has no effect whatever on the calculation. In fact you can turn your back each time the pickup is made! All you need know is the position *on the table* of the pile in which the chosen card falls after each deal. The same chart is used. Consider the first pile dealt as the "top," the next pile as "middle," and the last pile as "bottom." Add the key numbers after each designation of the pile, and the total will tell you the position of the chosen card in the *original order* of the 27. You can either memorize this order or have it written on a small card. If you turn your back each time the piles are assembled (and this will greatly add to the effect), it is a simple matter to refer to the small card while your back is turned.

It should be mentioned that all the effects described in this chapter can also be performed by dealing the cards face down rather than face up.   In fact this is how the trick is described by Professor Hoffmann.   If this procedure is followed, a minor alteration must be made in the chart.   The order of numbers in the second deal is reversed—reading (downward) 0, 3, 6 instead of 6, 3, 0.   The other numbers remain the same.   In Walker's method of performing the second effect, count 1, 2, 3 —1, 2, 3—etc., during the second deal, rather than 3, 2, 1.

If the piles are dealt face down, it is necessary for the spectator to pick up each pile to see if his card is there.   Or if you prefer, you can pick up each pile and fan it with the faces toward the audience.   This slows up the trick, but in some cases makes it seem more mysterious since at no time do you see the faces of the cards.

If the fanning procedure is used, a clever method of obtaining the name of the card is as follows   Place the designated pile in the middle during the first two pickups.   After the last deal, the chosen card will be exactly in the center of one of the three piles.   As you fan each pile, secretly bend back the lower corner of the center card with your left thumb.   The fan will conceal this maneuver from the audience, but you will be able to see the index of the card.   Therefore when the spectator tells you he sees his card in the fan, you immediately know the card's name as well as its position.

It is possible of course to perform the effects in this chapter by using all 52 cards in a pack, rather than just 27, but then it is necessary to deal the cards four times instead of three. See Jean Hugard's ENCYCLOPEDIA OF CARD TRICKS, p. 182, for a 52-card method of bringing the chosen card to any desired position in the deck.

Relation to Ternary System

Mel Stover, of Winnipeg, Canada, calls my attention to the application of the ternary counting system to the Gergonne pile trick. To make the application clear, let us first list the ternary equivalents of numbers from 0 through 27.

Decimal	Ternary	Decimal	Ternary	Decimal	Ternary	Decimal	Ternary
0	000	7	021	14	112	21	210
1	001	8	022	15	120	22	211
2	002	9	100	16	121	23	212
3	010	10	101	17	122	24	220
4	011	11	102	18	200	25	221
5	012	12	110	19	201	26	222
6	020	13	111	20	202	27	1000

The last digit of a ternary number indicates units, the next to last digit stands for "threes," the third from last stands for "nines," and so on. Thus to translate the ternary number 122 into our decimal system you have only to multiply the first digit by 9, add the product of the second digit times 3, then add the last digit. In this case, 9 plus 6 plus 2 equals 17—the decimal equivalent of the ternary number 122. Conversely, to find the ternary equivalent of 17 we first divide it by 9 (to obtain 1), divide the remainder (8) by 3 (to obtain 2), leaving us with a unit remainder of 2. Hence the ternary equivalent of 17 is 122.

To see how all this applies to the three-pile problem, let us suppose that you wish to bring the chosen card to position 19. To do this you must bring 18 cards above it. The ternary equivalent of 18 is 200. Taking these digits in reverse order gives us 002. This tells us how to pick up the piles each time —0 standing for top, 1 for middle, 2 for bottom. In other words, our first pickup places the pile containing the chosen card on top, the second assembly does the same, and the final assembly places the pile on the bottom. The card will then be 19th from the top.

Gargantua's Ten-pile Problem

Reflecting on the above matters led Mr. Stover to the invention of a truly stupendous breath-taking version of the trick. It makes use of the decimal system and a deck of 10 billion playing cards! The best way to obtain such a deck, writes

Stover (not seriously of course) is to buy 200 million decks of 52 cards each, then discard two cards from each deck. The spectator shuffles this Gargantuan pack, then while your back is turned he marks one card in such a way that the mark can be seen only by close inspection. After this is done, he gives you a number from 1 through 10 billion. By dealing the cards ten times, into ten piles of a billion cards each, assembling the piles after you have been told which pile contains the marked card, you are able to bring the card to the desired position.

Because the decimal system applies to this ten-pile version, it is a simple matter to determine the order in which the piles must be picked up during each of the ten assemblies. Let us assume the spectator wants the card to end up in position 8,072,489,392. Subtract 1 to obtain 8,072,489,391—the number of cards that must be brought above the marked one. Take the digits of this number in reverse order, remembering that 0 stands for top, 1 for second position, 2 for third position, and so on to 9 for bottom or tenth position. After ten pickups the marked card should be found at the 8,072,489,392nd position from the top.

"Care must be taken not to make a mistake in the dealing," Mr. Stover cautions in a letter, "as this would necessitate repeating the trick, and few spectators would care to see it a second time."

Chapter Four

Magic With Common Objects

Almost every common object that bears numbers has been exploited by magicians for purposes of mathematical magic. Tricks with playing cards, the largest category, were discussed in the previous chapters. In this chapter and the next we shall consider mathematical tricks with other common objects. Again, no attempt at an exhaustive treatment will be made, because the number of effects is too large, but I shall try to select those which are most entertaining and which illustrate the widest variety of principles.

DICE

Dice are as old as playing cards, and their origin equally obscure. It is surprising to learn that the earliest known dice, in ancient Greece, Egypt, and the Orient, were constructed exactly like modern dice—that is, with spots from one to six arranged on the faces of a cube in such manner that opposite sides total seven. Perhaps this is not so surprising when one considers the following facts. Only a regular polyhedron will insure equal odds for each face, and of the five regular polyhedrons, the cube has obvious advantages as a gaming device. It is the easiest to construct, and it is the only one of the five that rolls easily, but not too easily. (The tetrahedron and octahedron hardly roll at all, and the icosahedron and dodecahedron are so nearly spherical that they quickly roll out of reach.) Since a cube has six faces, the first six integers immediately suggest themselves, and the seven-arrangement provides a maximum of simplicity and symmetry. It is, of course, the only way the six figures can be paired so the sum of each pair is a constant.

It is this seven-principle which forms the basis of most mathematical tricks with dice. In the better tricks, however, the principle is so subtly employed that its use is not suspected. Consider, for example, the following trick, which is very old.

Guessing the Total

The magician turns his back while a spectator throws three dice on the table. He is instructed to add the faces. He then picks up any *one* die, adding the number on the *bottom* to the previous total. This same die is rolled again. The number it now shows is also added to the total. The magician turns around. He calls attention to the fact that he has no way of knowing which of the three cubes was used for the second roll. He picks up the dice, shakes them in his hand a moment, then correctly announces the final sum.

Method: Before the magician picks up the dice he totals their faces. Seven added to this number gives the total obtained by the spectator.

Frank Dodd's Prediction

Another clever trick using the seven-principle was contributed by Frank Dodd, of New York City, to *The Jinx*, September, 1937. The magician turns his back, instructing the spectator to form a stack of three dice, one on top of the other. He is told to add the two faces which are touching on the top and middle dice. To this sum he adds the sum of the touching faces on the middle and bottom dice. Finally, he adds the under side of the lower die to the previous total. The stack is covered with a hat.

The magician turns around and takes from his pocket a handful of matches. When the matches are counted, the number proves to be the same as the total obtained by adding the five faces.

Method: After the spectator has added the faces, the magician looks back over his shoulder for a moment to tell the spectator to cover the stack with a hat. When the magician does this, he glimpses the top face of the uppermost die. We will assume it is a six. In his pocket are 21 matches. He picks up all of them, but before removing his hand, allows six matches to drop back into the pocket. In other words, he removes all the matches except a number corresponding to the number on top of the stack. The number of matches in his hand will now correspond to the total of the five faces.

The fact that the spectator adds the faces which are touching, rather than opposite sides of the dice, serves to conceal the use made of the seven-principle. The handling of the matches was suggested by Gerald L. Kaufman, New York architect and author of THE BOOK OF MODERN PUZZLES.

Although the foregoing trick makes use of the seven-principle, actually it is possible to determine the hidden faces in a stack of dice simply by noting any two of the faces of each die. This is made possible by the fact that there are only two ways in which dice can be numbered, one a mirror-image of the other, and the fact that all modern dice are numbered on a "counterclockwise" basis. This means that if you hold a die so that you see only the 1, 2, 3 faces, the sequence of these numbers form a counterclockwise rotation (Fig. 4). By fixing these relative positions in mind, and recalling the seven-principle, it is possible to look at a stack of dice (the top face of the upper die covered by a coin) and name correctly the upper faces of each die. With a good visual imagination and a little practice, the feat can be performed with astonishing speed.

Positional Notation Tricks

Many interesting dice tricks involve positional notation. The following is typical. While the performer is not looking, the spectator throws three dice. The number of one die is multiplied by two, five is added, and the result multiplied by five. The upper face of a second die is added to the previous sum, then the result is multiplied by ten. Finally, the number on the remaining die is added. When the magician is given the end result, he quickly names the faces of the three dice.

Method: He subtracts 250. The three figures in the answer correspond to the faces of the three dice.

Hummer's Die Mystery

The popular parlor game of Twenty Questions, in which a group tries to guess what someone is thinking of by asking no more than twenty questions, is a good example of a principle that lies behind many mathematical tricks. We shall call it the Principle of Progressive Elimination. Gergonne's three-pile effect, explained in the previous chapter, determines the selected card by a progressive elimination of two-thirds of the cards until only a single card remains. Bob Hummer, in his booklet *Three Pets,* 1952, describes a dice trick involving a similar principle.

The effect of Hummer's trick is as follows. The performer sits at a table, but keeps his head turned to one side throughout the trick so that he cannot see the die at any time. Someone rolls the die and places it beneath the magician's cupped hands. A spectator is now asked to think of a number from 1 through 6. The performer raises his cupped hands so that the spectator sees the die. It has been placed so that three of its faces are visible, and the spectator is asked to state whether he does or does not see his number. The magician cups his hands over the die again and alters its position. He raises his hands and once more the spectator states whether he sees his number among the three visible faces. This is repeated a third time. The performer now covers the die and adjusts its position. When he raises his hands the selected number is uppermost on the die.

Method: A little reflection and you will see that three questions are quite sufficient to eliminate all but the chosen number. If you have a good visual imagination you should be able to do the trick immediately. The first question eliminates three of the faces. You must then turn the die so that two of the three possible faces are visible to the spectator, at the same time keeping track of the position of the third possible face. If the spectator does not see his number, you know immediately that it is the third face, so it is not necessary to ask a third question. If he does see his number, you know it is one of two faces, making it a simple matter to turn the die so that a final question will pinpoint the number.

The working of the trick is independent of the arrangement of numbers on the die, which means that the trick can be performed with a sugar cube. Let the spectator pencil a different number on each face or, if he prefers, he may draw letters or symbols. Still another presentation is to have the spectator mark only one face of the cube, leaving the other sides blank. In this case he simply tells you each time whether he sees the marked face, and you finish with the marked face uppermost. (For an elaboration of Hummer's trick, using a different principle of elimination, see Jack Yates' MINDS IN CLOSE-UP, 1954, published by Goodliffe, 15 Booth Street, Birmingham, England.)

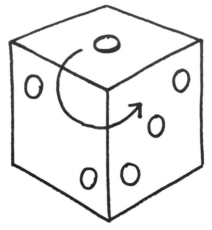

FIG. 4

DOMINOES

Dominoes have been used for mathematical magic much less than dice or cards. The following trick is the most widely known.

The Break in the Chain

The magician writes a prediction on a piece of paper which is folded and placed aside. The dominoes are mixed, then formed into a single chain by matching the ends as in regular play. After the chain is completed, the numbers at each end are noted. The paper is opened. On it are written the two numbers! The trick is repeated several times, the numbers being different with each repetition.

Method: The working depends on the fact that when a chain is formed out of a complete set of dominoes (usually 28 pieces), the end numbers always match. However, the magician secretly steals a domino before the trick begins, remembering its two numbers. These are the numbers he writes for his prediction. Since the full set forms an endless circuit, the missing domino will match the end numbers of the chain. To repeat the trick it is necessary for the magician surreptitiously to replace the stolen domino and remove a different one. In each case, the domino must be one with different numbers (not a double).

The Row of Thirteen

Another excellent domino trick employs thirteen dominoes placed face down in a row. While the magician is absent from the room, someone shifts any number of pieces from 1 to 12 from one end of the row to the other, moving them one at a time. The magician is called back. He immediately turns over a domino. The total of spots on its face indicates the number of pieces shifted. The trick can be repeated any number of times.

Method: The thirteen dominoes must bear spots which correspond in their totals to the first twelve integers. The thirteenth piece is a double blank. They are placed in a row, face down, in consecutive order, beginning with 1 on the left. The last domino on the right is the double blank. To illustrate how the pieces are to be moved, the magician shifts a few from the left end to the right. Before he leaves the room, he notes the number of spots on the domino on the *left* end. When he returns he counts mentally to the piece at this

position from the *right*.   For example, if the piece on the left
was a 6, he counts to the sixth domino from the right.   This
is the piece he turns over. If the left domino is blank, it is regarded as having the value of 13.

To repeat the trick it is a simple matter for the performer to
count to himself from the exposed piece to the one on the left
and ascertain its value before leaving the room again.

An amusing aspect of the trick is that if someone tries to
fool the magician by not moving any pieces, the domino turned
over will be the double blank.

CALENDARS

The arrangement of numbers on a calendar page has provided
material for many unusual tricks.   The following are among
the best.

Magic Squares

The magician turns away while a spectator selects a month
of the calendar, then draws on the page a square of such size
and in such position that it includes nine dates.   The magician
is told the smallest of these numbers.   After a moment's calculation he announces the sum of the nine numbers.

Method: Eight is added to the number and the result multi-
plied by nine.   For several other tricks similar to this, see
Tom Sellers' "Calendar Conjuring" on p. 117 of *Annemann's
Practical Mental Effects*, 1944.

Gibson's Circled Dates

A more complex trick, invented by the New York writer
Walter B. Gibson, appears on p. 119 of the book cited above.
The handling given here is slightly different from Gibson's,
and was worked out by the New York broker Royal V. Heath.
A stage presentation of the effect, by Magician Milbourne
Christopher, may be found in *Hugard's Magic Monthly*,
March, 1951.

The trick begins by letting a spectator select any monthly

page of the calendar. The magician turns his back and asks the spectator to circle at random one date in each of the five horizontal lines. (If the dates run into a sixth line, as they do on rare occasions, the sixth line is ignored.) The circled dates are then added to obtain a total.

With his back still turned, the magician asks, "How many Mondays did you circle?" This is followed by, "How many Tuesdays?" and so on through the days of the week. After the seventh and last question, the performer is able to give the sum of the five circled dates.

Method: The vertical column headed by the first day of the month has a key number of 75. Each successive column to the left (of course if the first day of the week is a Sunday there will be no columns to the left) has a key of five less. This enables the magician to glance at the calendar page before turning his back and quickly determine the key of the Sunday column. For example, if the first day of the month is Wednesday, then the Tuesday column has a key of 70, the Monday column, 65, and the Sunday column, 60. He remembers only the number 60—the key of the first column on the page.

Although the first question is about the number of Sundays circled, the magician ignores the answer completely. The number of circled Mondays is added to 60. The number of circled Tuesdays is multiplied by two and added to the previous sum. Wednesdays are multiplied by three and added to the running total. Thursdays by four, Fridays by five, and Saturdays by six. (The fingers may be used for keeping track of these six integers.) The final total will be the sum of the circled numbers.

Stover's Prediction

Another ingenious calendar trick, the invention of Mel Stover, appears as follows. The spectator draws on any calendar page a square enclosing sixteen dates. The magician glances at the square, then writes a prediction. Four numbers on the square are now selected by the spectator, apparently at random in the following manner. He first circles any one of the sixteen dates. The horizontal and vertical rows that contain this date are then crossed out by drawing lines through them. The spectator now circles any one of the remaining dates—that is, one of the dates not crossed out. Again, the horizontal and vertical rows containing this date are eliminated. A third date is selected in the same manner and its two rows crossed out. All the dates will now be crossed out except one. This last remaining date is circled. The four circled numbers are added. The sum is exactly what was predicted by the performer.

Method: The magician glances at two corner numbers, diagonally opposite each other. It does not matter which pair is used. These numbers are added and the sum doubled to obtain the answer.

A simple application of the principle, not requiring a calendar, is to draw a checkerboard of sixteen squares and number them from 1 to 16, in normal reading order. The spectator chooses four numbers, by the process described above, and adds them. The total will be 34 in all cases. The principle can be applied of course to squares of any desired size.

Calendar Memorizing

A popular feat among professional memory experts is that of giving quickly the day of the week for any date named by someone in the audience. The trick is accomplished by a complex calculation that can be greatly speeded up by the use of mnemonic devices. Excellent methods are described in Bernard Zufall's CALENDAR MEMORIZING, 1940 (No. 3 of a series of booklets titled ZUFALL'S MEMORY TRIX), and in Wallace Lee's MATH MIRACLES, 1950. The effect is discussed in many books on mathematical recreations as well as in books on mnemonics.

WATCHES

Tapping the Hours

One of the oldest tricks in magic is performed with a watch (or clock) and a pencil. The spectator is asked to think of any number on the dial. The magician starts tapping numbers with the pencil, apparently at random. As he taps, the spec-

tator counts silently, beginning with his number on the first
tap. When he reaches twenty he calls out "Stop." Oddly
enough, the magician's pencil at this moment is resting on the
original number mentally selected.

Method: The first eight taps are made at random. The
ninth tap is on 12. From this point on, the numbers are
tapped counterclockwise from 12. When the spectator calls
"Stop," the pencil will be resting on the chosen number.

Instead of telling the spectator to stop you when he reaches
20 in his silent count, you may permit him to do this at any
number higher than 12. He must of course tell you the number at which he intends to call out "Stop." Simply subtract
12 from this number. The remainder tells you how many
taps you make at random before you start tapping counterclockwise from 12 o'clock.

The tapping principle employed here has been applied to
dozens of other effects, some of which will be mentioned in
the sixth chapter. Eddie Joseph, in his booklet TRICKS FOR
INFORMAL OCCASIONS, describes an effect with 16 blank cards or
slips of paper, the operation of which is the same as the watch
effect. Sixteen words are called out by the audience. Each
word is written on a blank card, then the backs are lettered
from A to P. The cards are mixed on the table. The
magician turns his back, someone selects a card, notes both
word and letter, then shuffles it back with the others. The
magician now picks up the cards and fans them in his hand,
the word sides facing the audience. He appears to pick cards
at random, tossing them one at a time on the table, while the
spectator recites to himself the letters of the alphabet, beginning with the letter on his chosen card. When he reaches P
he calls out "Stop." The card which the performer is about
to toss down proves to be the selected one.

To work the trick, simply toss the cards in reverse alphabetical order, beginning with P.

Die and Watch Mystery

Another watch trick, of my own devising, is as follows.
While the performer's back is turned, a spectator rolls a die.

He then thinks of any number, preferably under 50 to make the trick faster. We will assume he chooses 19. Beginning with the number on the watch dial indicated by the die, he starts tapping clockwise, counting his taps until he gets to 19. The number reached by the 19th tap is written down. He then goes back to the starting point (the number on the die) and follows the same procedure, but tapping counterclockwise. Again, the number at the 19th tap is written down. These two numbers are added. The total is called out. Immediately, the magician announces the number on the die.

Method: If the number called is under 12, it is halved to obtain the answer. If it is over 12, the performer subtracts 12, then halves the result.

DOLLAR BILLS

Heath's Bill Trick

For more than twenty-five years Royal V. Heath has been performing an interesting trick using the serial number on a dollar bill. The effect is as follows. A spectator takes a bill from his pocket and holds it so the performer cannot see its serial number. He is asked to state the sum of the first and second digits, then the sum of the second and third, the third and fourth, and so on until the end of the eight-digit number is reached One additional total is asked for, that of the last and second digits. As these totals are called, the magician jots them down on a piece of paper. After a brief mental calculation, he is able to call out the original serial number. The formula by which the number is calculated is published here with Mr. Heath's permission.

Method: As the spectator calls out the sums of the paired digits, jot them down in a row from left to right. When the end of the serial number is reached, you will have written seven sums. As you write them down, add in your mind the second, fourth, and sixth sums, remembering the grand total. Now ask for the sum of the last and second digit, and add this to the previous total (the reason for adding the first three sums in advance is that it speeds up the final calculation). You now have in mind the total of alternate numbers (beginning with the second) in your row of eight sums.

The next step is to add in your mind the third, fifth, and seventh numbers in your row of sums. Subtract this total from the first total and divide the remainder by two. The answer will be the *second* digit of the original serial number. It is now a simple matter to obtain the other digits. Subtracting the second digit from the first sum gives you the first digit of the original serial number. Subtracting the second digit from the second sum gives you the third digit. The third digit from the third sum gives the fourth digit, and so on to the end, ignoring the final sum. A sample serial number diagrammed in Fig. 5, should make this clear.

This method applies to any number containing an even number of digits and is therefore the method used on dollar bill serial numbers, which always contain eight digits. However, you may wish to perform the same trick with some other series of numbers that may contain an odd number of digits, such as a person's phone number or license number. In this case a slightly different procedure and formula must be used.

Instead of asking for the sum of the last and second digits, after the end of the number is reached, ask instead for the sum of the last and *first*. To learn the original number you must first obtain the total of all alternate numbers (in your row of sums) beginning with the *first* sum instead of the second. From this total you subtract the total of all remaining alternate sums. The remainder divided by two gives you the *first* (rather than the second) digit of the spectator's original number. It is now easy to obtain the other digits. Fig. 6 shows the procedure to be followed, assuming that the spectator's phone number is 3–1–1–0–7.

It is not necessary for the numbers in the original series to be digits. Consequently you may ask someone to jot down a list of numbers, making each number as large or as small as he wishes. You need not ask whether there is an even or an odd number of these numbers, because you will know this by the time he has finished calling off the sums of the pairs. You then ask for the sum of last and first number, or the sum of

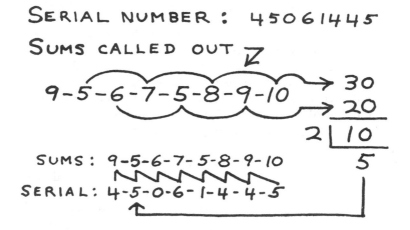

FIG. 5

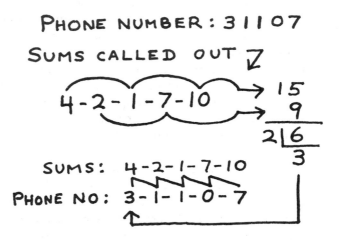

FIG. 6

last and second number (depending on whether the series of original numbers is odd or even), and proceed in the manner already described.

MATCHES

Many mathematical tricks have been devised which employ small objects as counting units. I have already described a few of this nature in the first chapter. The following effects lend themselves more readily to matches, though of course other small objects—coins, pebbles, or bits of paper—could be used.

The Three Heaps

The magician faces away from the audience while someone forms three heaps of matches on the table. Any number of matches may be used provided the heaps are equal, and there are more than three matches in each. The spectator calls out any number from 1 to 12. Although the magician does not know the number in each heap, he is able to give directions for shifting the matches so that the number in the center heap is equal to the number named.

Method: The spectator is asked to take three matches from each of the two end heaps and add them to the center. Then he counts the number in either end heap and removes that number from the center, placing them on either end. This procedure always leaves nine matches in the middle heap, so it is an easy matter to give instructions for bringing this heap to the desired total.

Match Folder Mind-reading

A trick using a similar principle was contributed by Frederick DeMuth to an early issue of *The Jinx*. A new match folder (containing twenty matches) must be used. The magician requests that while his back is turned the spectator tear out a few matches (the number must be less than ten) and place them in his pocket. He then counts the number remaining in the folder. We will assume it is 14. Enough additional matches are torn from the folder to form "14" on the table. The "14" is formed by placing one match to the left to represent the first digit, then four matches to the right to represent the second. These matches are picked up and also placed in the pocket. Finally, the spectator tears out a few more matches and holds them in his closed fist. The magician turns around, takes a quick look at the folder, and states the number of matches the spectator is holding.

Method: The number of matches left in the folder is subtracted from nine.

The Tramps and Chickens

Employing a radically different principle is an old match stunt usually presented in the form of a story about two tramps and five chickens. The magician places five matches on the table to indicate chickens. A single match in each hand represents the tramps. They steal the chickens by grabbing them one at a time. (The five matches are picked up one at a time, alternating hands.) But the tramps hear the farmer walking toward them, so they replace the chickens. (One by one the five matches are returned to the table.) After the farmer has passed by, without discovering the thieves who have hidden behind bushes, they sneak up to the chickens and grab them again. (The matches are picked up as before.) At this point one of the tramps begins to complain. For some curious reason he has only one chicken (the left hand opens to show only two matches) while the other tramp has four (the right hand opens to reveal five matches).

Method: When the matches are first picked up, the *right* hand takes the first match. When replaced, the *left* hand puts down the first match. This leaves the left hand empty, but the performer keeps it closed as though it still contained a match. He begins with the *right* hand when he picks up the five again. This puts five matches in the right hand and two in the left.

The Purloined Objects

Another ancient match trick makes use of 24 matches placed
on the table beside three small objects—say a penny, ring, and
key.   Three spectators, whom we shall call 1, 2, and 3, are
asked to take part.   Spectator 1 is given one match.   Spectator
2 gets two matches.   Spectator 3 gets three matches.   You now
turn your back and ask each spectator to take one of the three
objects.   We shall designate the objects as A, B, and C.

Tell the person who took A to remove from the pile of
matches as many as he already holds.   Tell the person who
took B to remove twice as many as he holds.   Tell the remaining spectator, who took C, to remove four times as many
matches as he already possesses.   All three spectators put their
matches and their object in their pocket.

You now turn around, glance at the number of remaining
matches, and immediately tell each person which object he is
holding.

Method: If 1 match remains, Spectators 1, 2, and 3 hold
objects A, B, and C in that order.

If 2 remain, the order of objects is B, A, C.

If 3 remain, the order is A, C, B.

If 4 remain, then someone made an error, as it is impossible
for this to happen.

If 5 remain, the order is B, C, A.

If 6 remain, the order is C, A, B.

If 7 remain, the order is C, B, A.

The trick is explained in several medieval treatises on
mathematical recreations.   For a recent discussion see Ball's
MATHEMATICAL RECREATIONS, revised American edition, 1947,
p. 30ff. Mnemonic devices for performing several variations of
the effect may be found on pages 23 and 218 of THE MAGICIAN'S
OWN BOOK.   Since 1900 many versions of the trick, with different memory devices, have been published in books or sold
as individual tricks.

The most common mnemonic device is a list of words in
which certain consonants stand for the three objects.   For example, Clyde Cairy, of East Lansing, Michigan, proposes doing

the trick with a toothpick, lipstick, and ring. The following words must be memorized:

1	2	3	5	6	7
TAILOR	ALTAR	TRAIL	ALERT	RATTLE	RELATE

The letter T stands for toothpick, L for lipstick, and R for ring. All three letters appear in each word in an order which corresponds to the order of objects. The numbers above each word indicate the number of remaining matches.

In 1951 George Blake, of Leeds, England, published a clever handling of the trick in which the remaining matches are concealed in a match box while the magician's back is still turned. His manuscript, *The ABC Triple Divination*, must be consulted for his clever sleight-of-hand method for learning the number of matches in the closed match box. We are concerned here only with his simplified formula. The objects are designated by A, B, and C, and the following sentence memorized:

1	2	3	(4)	5	6	7
ABIE'S	BANK	ACCOUNT	(SOON)	BECOMES	CASH	CLUB

You will note that each word contains only a pair of the key letters, but in each case the missing letter follows the pair. Thus the first word gives you AB, to which you add the missing letter to make ABC. The fourth word, SOON, is included only to make each word correspond to the number of remaining matches, thereby simplifying the task of counting to the desired word. It is not possible, of course, to have four matches remaining, consequently SOON contains no key letters.

Oscar Weigle uses a similar type of sentence. He divides the objects by their size into small, medium, and large. The initial letters of these words, S, M, and L, are the key letters in the following sentence:

1	2	3	(4)	5	6	7
SAM	MOVES	SLOWLY	(SINCE)	MULE	LOST	LIMB

As in Blake's sentence, each word contains two key letters, to which the missing letter is then added.

COINS

Coins possess three properties that may be exploited in mathematical tricks. They can be used as counting units, they possess numerical values, and they have "head" and "tail" sides. Here are three effects, each of which illustrates one of the three properties.

The Nine Mystery

A dozen or more coins are placed on the table in the form of a figure nine (Fig. 7). While the magician's back is turned someone thinks of any number greater than the number of coins in the tail of the nine. He starts counting from the end of the tail, counting up and around the nine counterclockwise until he reaches the number. Then he starts counting at 1 again, beginning on the last coin touched, but this time he counts clockwise around the circle, circling it until he reaches the selected number again. A tiny piece of paper is concealed beneath the coin on which his count terminates. The magician turns around and immediately lifts this coin.

Method: Regardless of the number chosen, the count will always end on the same coin. Experiment with any number first, making the count mentally, to learn which coin this will be. If you repeat the trick, add more coins to the tail so that the count will fall on a different coin.

Which Hand?

An ancient trick using the values of coins is performed as follows. Ask someone to hold a dime in one fist, a penny in his other fist. Tell him to multiply the value of the coin in his right hand by eight (or any even number you wish to use) and the value of the other coin by five (or any odd number you care to name). He adds the results and tells you whether

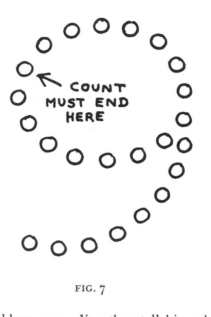

COUNT MUST END HERE

FIG. 7

the answer is odd or even. You then tell him which coin is in which hand.

Method: If the number he gives you is even, his right hand holds the penny. If odd, it holds the dime.

Heath's Variation

An amusing variation of this trick is given by Royal V. Heath in his book MATHEMAGIC. In this version the spectator holds a penny in one fist, a nickel in the other. Ask him to multiply the coin in his left hand by fourteen. After he has done this, tell him to do the same with the other coin. He then adds the two results and gives you the total. You meditate a moment, as though performing difficult calculations in your head, and tell him which hand holds which coin.

Method: The total he gives you has nothing whatever to do with the trick. You merely note which hand causes him to take the longer time for his mental multiplication. It will naturally be the hand holding the nickel.

Heads or Tails?

An interesting trick using the two sides of coins begins by placing a handful of change on the table. The magician turns his back and asks someone to start turning the coins over one at a time, picking coins at random. Each time he turns a coin he calls out "Turn." He continues doing this as long as he wishes and the same coin may be turned as often as desired. He then covers one coin with his hand. The performer turns around and correctly states whether the covered coin is heads or tails.

Method: Before turning your back count the number of heads showing. Each time he calls out "Turn," add one to this number. If the final total is even, there will be an even number of heads showing at the finish. If odd, an odd number of heads. By examining the uncovered coins it is a simple matter to determine whether the concealed coin is heads or tails.

The trick can be performed with any group of objects that can be placed in one of two possible positions—soda bottle tops, pieces of paper with an "X" on one side, playing cards, match folders, and so on.

A more complex variation of this trick appears in Walter Gibson's PROFESSIONAL MAGIC FOR AMATEURS, 1947. Three pieces of cardboard are used. A colored spot is on the front and back of each piece, making six spots altogether, each spot a different color. The trick is presented exactly as with the coins. To calculate the color showing on the covered card, regard each card as having a "head" and "tail." For example, the three primary colors—red, yellow, and blue—may be considered "heads" and the three secondary colors—green, orange, and purple—may be considered "tails." It is also necessary to recall how the colors are paired on each card. This is easily remembered if the pairs are complementary colors—i.e., red and green, blue and orange, purple and yellow. By knowing the number of "head" colors that should be uppermost at the finish, it is an easy matter to calculate the color uppermost on the concealed card.

Words, letters, numbers, symbols, etc., may of course be substituted for colors on the three cards. If you are skilled in mnemonics, have the spectator call out any six words which you write on the six faces of the cards. It is necessary to memorize the words in pairs and also to designate one member of each pair as the "head" and the other word as the "tail." Or more simply, assign the value of 1 to three of the words, and 0 to their respective partners.

Bob Hummer, in his manuscript *Three Pets*, describes a trick with two nickels and three pennies that makes use of the same turning principle. His poker chip trick, to be explained in Chapter Six, also operates on a similar principle, and so does a poker chip trick contributed by Dr. L. Vosburgh Lyons, of New York City, to *The Phoenix*, May 1, 1942.

CHECKERBOARDS

Hummer's Checker Trick

Bob Hummer was the first, I believe, to devise a mathematical trick making use of a checkerboard. The trick was marketed under the name of "Politicians Puzzle," and is included here with Hummer's permission. The marketed version used a miniature cardboard board with six squares to the side, but the effect is easily extended to a regulation-size board.

A spectator is given three checkers. While the performer's back is turned he places them either in the corner row indicated by the three A's in Fig. 8, or the opposite corner row marked with three B's. He now spells to himself the letters in his name, moving any checker he wishes for each letter. The moves are made in any direction as though the checkers were kings. After the name has been spelled, he then spells it again, moving the checkers at random as before. He continues doing this as long as he wishes, but must stop at the termination of one of the spellings. You then turn around, glance at the board, and tell him whether he started with his checkers in the upper left corner or the lower right.

Method: The name to be spelled must have an even number of letters. If the spectator's first and last names are even, he may choose either name to spell. If one is even and the other odd, request that he spell the even one. If both are odd, he must spell his entire name (since the total of two odd numbers is always even).

When you turn around, examine the even vertical rows, regarding the rows as numbered from left to right as shown. If you find an even number of checkers in these rows (zero is considered even), then you know he started from the lower right corner. Otherwise, the upper left corner. Once the principle is understood, other variations will suggest themselves.

MISCELLANEOUS OBJECTS

Hummer's Three-object Divination

A beautifully conceived trick, using three small objects, was marketed by Bob Hummer in 1951 under the title "Mathematical Three-Card Monte." Although Hummer's description involved three playing cards, the effect is applicable to any three objects. It is described here with Hummer's permission.

The three objects are placed in a row on the table, their positions designated as 1, 2, 3. These numbers refer not to the objects, but to their *positions*. The magician turns his back. The spectator now switches the position of any two objects, calling out the numbers of the two positions. Thus if he switches the objects in positions 1 and 3 he calls out the numbers 1 and 3. He continues switching pairs of objects as long as he wishes, always calling out the numbers with each change. When he tires of doing this, he pauses and thinks of any one of the three objects. Then he switches the position of the other two *without telling the magician the positions involved in this switch*. After this is done, he goes back to switching pairs at random again, each time calling out the positions being switched, and continues until he tires. The magician turns around and immediately points to the chosen object.

Method: While your back is turned, you make use of one hand as a calculating device. Three of your fingers are designated as 1, 2, and 3. Before turning your back, note the

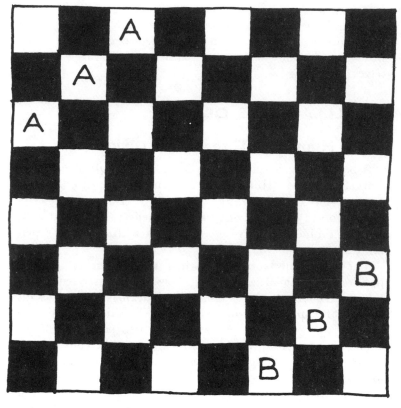

FIG. 8

position of any one object.   Let us assume the three objects are a ring, pencil, and coin, and that you observe that the ring is in position 1.   Place your thumb against the tip of the finger you are calling 1.

As the spectator calls out the switches to you, your thumb moves from finger to finger following the position of the ring only.   Thus if the first switch involves 1 and 3, you move your thumb to finger 3.   But if the switch involves 2 and 3, the ring is not involved, so you do nothing, keeping your thumb on finger 1.

After the spectator has thought of an object and made his secret switch of the other two, he begins calling switches again. *You continue to follow the ring, even though his secret switch may have changed the ring's position.*

At the conclusion of the switching, your thumb will be resting against a finger.   Let us say it is touching finger number 2. Glance at position 2 on the table.   If the ring is in this position, you know immediately that he thought of the ring, because its position has remained unchanged throughout the trick.

If the ring is not in the position indicated by your thumb, then glance at the other two objects.   They will naturally be the ring and another object.   The object which is *not* the ring will be the one thought of.

The method is delightfully simple.   If you play with the effect for a while you will soon discover why it works. Actually, the problem is one of elementary logic, with the fingers used as a simple logic machine.

The trick is extremely effective when done with three face-down cards.   It is only necessary for you to have a secret mark on the back of one card, such as a pencil dot or a slight crimp of one corner.   This is the card that you follow with your fingers while your back is turned.   When it is time for the spectator to think of one of the cards, he must of course peek at its face and remember it.   When you turn around you can immediately turn over the chosen card, even though the faces of the cards have been concealed from you throughout the trick.

An excellent dinner table variation is to let the spectator, while your back is turned, conceal a match folder under one of three inverted coffee cups. Ask him first to switch the positions of the two empty cups without telling you which cups they are. He then switches cups at random by sliding them along the table, each time calling out the two positions involved. You turn around and immediately lift the cup covering the folder, even though the spectator himself may have lost track of where the folder is. To do this, it is necessary to find some small flaw or identifying mark on one of the cups so that you can follow this cup with your fingers as previously explained.

If spectators are in position to observe you while your back is turned, put your hands in your pockets. In this way the use of your fingers as a calculating device will not be noticeable.

Yates' Four-Object Divination

Jack Yates, in MINDS IN CLOSE-UP (previously cited), describes an intriguing trick with four matches. The trick was suggested to him by the Hummer trick just explained, and appears as follows. Four matches are placed side by side in a row on the table. Three of the matches point in one direction, one match in the opposite direction, to distinguish it from the others. While the magician's back is turned, a spectator shifts the matches about in an apparently random manner. With his back still turned, the magician asks the spectator to remove first one match, then another, and then another, leaving only one match on the table. The remaining match is the reversed one.

The trick may be repeated many times, but always with the same results. Since any four objects may be used, the trick is included here rather than under the previous sub-head of matches.

Method: The four matches or objects are placed on the table in positions which we shall designate as 1, 2, 3, and 4. Ask someone to name one of the objects. Note its position before you turn your back. Now ask the spectator to make five switches, each time exchanging the chosen object with an object *adjacent* to it. If the chosen object is at one end of the row then of course there is only one move that can be made, but if it is not on the end it can be switched with the object on either its right or left.

Since the spectator tells you nothing whatever about the nature of these five switches, one would think the chosen object might be brought to any position in the row. This is not, however, the case. If the original position of the object was 2 or 4 (even numbers) it will end at positions 1 or 3 (odd numbers). Conversely, if the starting position was 1 or 3, it will end at 2 or 4. This result always obtains if an odd number of switches are made. In the example given here, you specify five switches, but you could just as well allow seven or twenty-nine or any other odd number. If you prefer, you may allow an *even* number of switches, but in this case the object ends at an even position if it starts from an even one, or an odd position if it starts from an odd one. Thus you may let the spectator himself decide each time how many switches he intends to make, provided of course he tells you the number. Another presentation is to let him spell the letters of his name as he makes the exchanges.

After the switches are concluded, you must then direct the spectator in the removal of three objects, one at a time, leaving the chosen object on the table. This is done as follows:

If the object's final position is known to you to be 1 or 3, ask him to take away the object at position 4. You can do this simply by waving the proper hand and saying, "Please take away the object at *this* end." After this is done, ask him to switch the selected object once more with an adjacent one. *This final switch always puts the object in the center of the three that remain!* It is now a simple matter for you to direct the removal of the next two objects so that only the chosen one is left.

If, on the other hand, the object's final position is 2 or 4, then ask the spectator to take away the object at position 1. Now ask him to switch the object with an adjacent one. As before, this will bring the selected object to the center of the remaining three, making it easy for you to specify the next two removals.

A slightly different presentation of the first part of the trick is to let the spectator make as many switches as he wishes, keeping this up as long as he desires and stopping whenever he feels like it. In this case he must call out "switch" each time he makes an exchange. As in the previous trick of Hummer's, you may use your fingers as a simple counting device to keep track of his switches. Call the index finger 1 and the middle finger 2. If the object's original position is odd, put your thumb against the index or "odd" finger. If even, put your thumb against the middle or "even" finger. Now as the switches are called, keep changing your thumb from finger to finger. At the conclusion, if your thumb is on the "odd" finger, you know the object is at position 1 or 3. If the thumb is on the "even" finger, the object is at 2 or 4.

Mel Stover suggests that the trick be performed with four empty glasses and one ice cube. The cube is poured from glass to adjacent glass. In this version the spectator may remain silent as it is easy for you to keep track of the highly audible clinks.

Chapter Five

TOPOLOGICAL TOMFOOLERY

In previous chapters we have considered only tricks that are mathematical in their method of operation. We have not included magic in which the effect alone is mathematical. For example, a magician may deal four perfect bridge hands from a previously shuffled deck. This is mathematical in the sense that a disordered series is transformed mysteriously into an ordered one. But if the operation depends not on mathematics but simply on a secret exchange of one pack for another, it would not be regarded as a mathematical trick.

A similar approach is taken in this chapter. A great many magic tricks can, in a broad sense, be called topological because they seem to violate elementary topological laws. One of the oldest tricks in magic, known as the "Chinese Linking Rings," is in this category. Six or more large steel rings are caused to link and unlink mysteriously—feats obviously impossible in view of the properties of simple closed curves. Tricks in which rings are removed from, or placed upon, a cord or stick held firmly at both ends by a spectator can be regarded as magical linking and unlinking, because the cord and spectator form a closed circuit through which the ring is linked. But most of these tricks with rings operate by mechanical methods, sleight-of-hand, or other magical means that have nothing to do with topology.

Much closer to what might be called a topological trick is an effect usually sold by magic dealers under the name of "Tumble Rings." This is a chain of rings, linked together in an exceedingly odd way. By manipulating it properly, a ring at the top of the chain seems to "tumble" down the chain, finally linking itself mysteriously to the lowest ring. The

trick is self-working and there is no question but that the
linked rings form a complex and curious topological structure.
But the "tumbling" is merely an optical illusion, created
mechanically and not involving topological laws.

The material that follows, therefore, will deal only with
tricks in which the mode of operation may be considered
topological. As might be expected, in view of the fact that
topology is concerned with properties invariant under con-
tinuous transformation of the object, the field of topological
magic is restricted almost entirely to such flexible material as
paper, cloth, string, rope, and rubber bands.

The Afghan Bands

The familiar "Moebius strip," named for Augustus Ferdi-
nand Moebius, the German astronomer and pioneer topologist
who first described the surface, has been exploited by magicians
for at least seventy-five years. The earliest reference I have
found to its use as a parlor trick is in the 1882 enlarged English edition of Gaston Tissandier's LES RECREATIONS SCIENTI-
FIQUES, first published in Paris in 1881. In this version the
magician hands a spectator three large paper bands, each
formed by pasting together the ends of a long paper strip.
With a pair of scissors the spectator cuts the first strip in half
lengthwise, cutting around the strip until he returns to the
starting point. The cutting results in two paper bands.
However, when he cuts the second strip in a similar manner,
he finds to his surprise that it forms one single band with
twice the circumference of the original. Cutting the third
strip produces an equally startling result—two paper rings
which are interlocked.

The working of the trick depends upon the preparation of
the bands. The first strip is joined end to end without twisting. The second band is a Moebius surface produced by
giving the strip a single twist before the ends are pasted. One
of the many curious properties of this surface, which has only
one side and edge, is that cutting it lengthwise results in a
single large ring. (If the cut is begun a third of the way from
the edge, instead of the center, the result is one large band

with a smaller band interlocked.) The third band is formed by giving the strip two twists before joining the ends.

The term "Afghan Bands" was applied to this paper version as early as 1904, when Professor Hoffmann called it by this name in LATER MAGIC. Why this curious name was given to the trick remains a mystery.

A later version of the same trick, which adds considerable comedy to the presentation, was developed by magician Phil Foxwell. The performer exhibits three huge paper bands prepared as previously described from strips of brown wrapping paper, each about 8 inches wide and 12 feet long. In this larger size the twists will not be as noticeable. Two people are called from the audience and each handed a band and pair of scissors. The performer exhibits a ten dollar bill, announcing that he will award it to the first person who succeeds in cutting his band into two separate rings. To illustrate what he means, the magician cuts the remaining band in half, showing the two rings.

The contestants begin at the command "Go!" As soon as they finish the magician starts to hand the ten dollar prize to the winner, only to discover that he failed to comply with instructions, having produced either a single ring or two interlocked. The prize is then offered to the other contestant, but it is soon apparent that he likewise was unable to produce separate bands.

About 1920 Carl Brema, an American magician, began doing the trick with red muslin instead of paper. The cloth bands could be ripped down the middle, making the trick more colorful and faster to perform. In England at about the same time, Ted Beal thought of presenting a paper version which began with a single large band. It was cut in half to form two paper bands, one with two twists and the other without twists. Each band was in turn cut in half by two spectators, one obtaining two separate rings and the other two linked rings. Beal's handling was explained in MORE COLLECTED MAGIC, by Percy Naldrett, published in 1921.

In America, a Philadelphia attorney and amateur magician, James C. Wobensmith, unaware of Beal's method, developed a method using a wide muslin band that was ripped in half to form two separate bands. When one band was torn it produced two linked rings. The other band was then torn to form a single large ring. Wobensmith's version was marketed by Brema, the first advertisement for it appearing in *The Sphinx*, January, 1922, and explained by Wobensmith in an article titled "The Red Muslin Band Trick," in *The Magic World*, Sept., 1923. Wobensmith's original method of preparing the band is pictured in Fig. 9. It was subsequently improved to the form in which it is now sold by dealers (Fig. 10). A quick-drying cement can be used for fastening the ends.

Harry Blackstone and S. S. Henry were the two most prominent professional magicians to feature the Wobensmith trick in their stage performances. It was presented with patter about a magician in a carnival side-show who was called upon to produce belts for the fat lady and the Siamese twins. After the original red muslin band is ripped in half to form two bands, one is then ripped to produce two linked bands for the twins and the other is ripped to make the large belt for the fat lady. This presentation first appeared in THE L.W. MYSTERIES FOR CHILDREN, 1928, by William Larsen and T. Page Wright.

In 1926 James A. Nelson contributed to *The Sphinx* (December issue) a method for preparing a paper band so that two cuttings produce a chain of three interlocked bands (Fig. 11). Ellis Stanyon published a pamphlet in London in 1930, titled REMARKABLE EVOLUTION OF THE AFGHAN BANDS, which gave fifteen paper variations. The pamphlet was reprinted in GREATER MAGIC, 1938, by John Hilliard. Jean Hugard's ANNUAL OF MAGIC, 1938–39, contains Lester Grimes' method of cutting a paper strip into a chain of five bands, an effect described in fuller detail in the October, 1949, *Magic Wand*.

In *Hugard's Magic Monthly*, Dec., 1949, I described two interesting muslin variations. One of them, originated by William R. Williston, is prepared as shown in Fig. 12. The first tear produces a large band twice the original size, and the second tear results in a still larger band, four times the first one. The second variation, which I worked out, is pictured in Fig. 13. The first tear gives a single large band and the second, two interlocked rings.

Many other combinations can be worked out. Woben-smith's present handling uses the band shown in Fig. 14. The first rip produces two separate bands. One is now torn to produce a chain of three linked bands. The other is torn to form a single large band. This band is then ripped once more to produce a still larger one.

Stanley Collins pointed out in the 1948–49 MAGIC WAND YEAR BOOK that if a small solid ring is placed on a strip, and the ends joined after three twists, the usual cutting or tearing will result in one large strip knotted around the ring.

HANDKERCHIEF TRICKS

Finger Escape

More than a dozen novel handkerchief tricks may be considered topological. A version of one of the oldest is as follows. The magician holds a handkerchief by opposite corners and swings it in circles, skip-rope fashion, to form a twisted cloth rope. This is placed over the extended right index finger of a spectator as shown in Fig. 15. After wrapping it about the finger, the spectator places his left index finger on top of the right one, and the cloth is twisted securely about both fingers. The performer now grasps the tip of the lower index finger, as shown in Fig. 22. The spectator is told to remove his other finger from the cloth. When the magician lifts up on the handkerchief, it pulls free of the finger he is holding.

Although the cloth appears to be tightly wrapped about both fingers, the method of wrapping is such that it leaves the spectator's right index finger *outside* the closed curve formed by the handkerchief. The method of wrapping is as follows:

1. Cross the handkerchief underneath the finger (Fig. 16). Note that the end marked *A* is toward you at the point of crossing. Throughout the rest of the moves this same end must always be toward you when the ends cross. Otherwise the trick will fail.

2. Cross the ends above (Fig. 17).

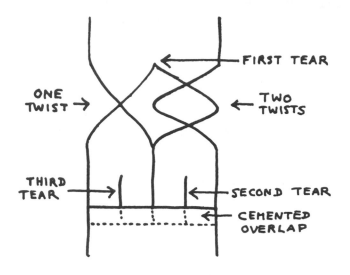

FIG. 9

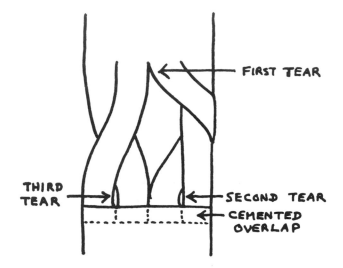

FIG. 10

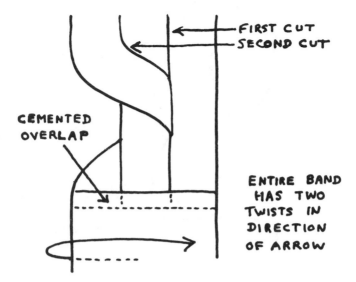

FIG. 11

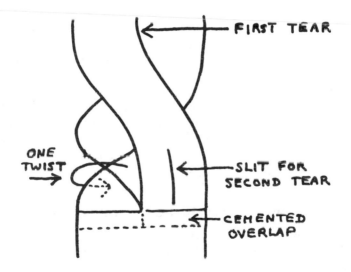

FIG. 12

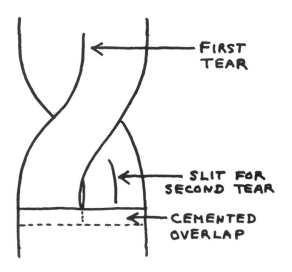

FIG. 13

3. The spectator places his left index finger on top of the crossing (Fig. 18).

4. Cross the ends above, taking care to keep the proper end toward you (Fig. 19).

5. Cross the ends beneath (Fig. 20).

6. Bring the ends up and hold them in the left hand (Fig. 21). The two fingers now appear securely wrapped together.

7. Grasp the tip of the lower finger. He removes the other finger from the cloth. Lift with the left hand. The cloth pulls free (Fig. 22).

Tabor's Interlocked Handkerchiefs

A similar trick was invented several years ago by Edwin Tabor, a magician in Berkeley, California. Two handkerchiefs, preferably of contrasting colors, are each twisted rope fashion. They are held in the left hand as shown in Fig. 23. The right hand reaches beneath the dark handkerchief,

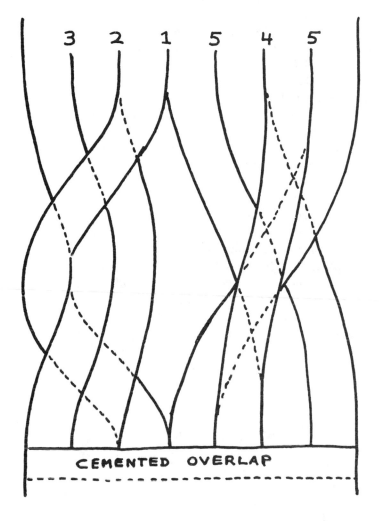

FIG. 14

grasps end A, and wraps it once around the other handkerchief (Fig. 24).

End B of the dark handkerchief is carried under and then over the other one, as indicated in Fig. 25.

Ends B and C are brought together below and held in the right hand. Ends A and D are brought together above and held in the left (Fig. 26).

The handkerchiefs seem to be tightly interlocked, but when the ends are pulled, they come apart easily. If large silks are used, each can be wrapped *twice* around the other, and still they come apart.

Both of the above tricks operate on the principle of having one series of wraps undo, so to speak, what has been done by another series. The same principle has also been applied to several rope tricks in which the rope is twisted about the leg, a post, or a stick, then pulled free.

Stewart Judah marketed an effect in 1950 under the title of "Judah Pencil, Straw, and Shoestring" in which the wrapping principle is cleverly applied. A shoestring is wrapped securely around a pencil and soda straw, placed side by side. When the string is pulled free it apparently penetrates the pencil, cutting the soda straw neatly in half. The trick is currently being sold by U. F. Grant, a magic dealer in Columbus, Ohio. A similar effect using a pencil, ring, string, and cigarette was worked out by Eddie Joseph and sold by the Abbott Magic Co., Colon, Mich., in 1952 under the title of "Ring Tie."

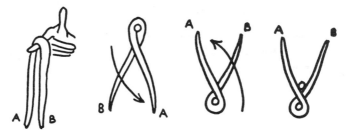

FIG. 15   FIG. 16   FIG. 17   FIG. 18

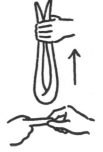

FIG. 19   FIG. 20   FIG. 21   FIG. 22

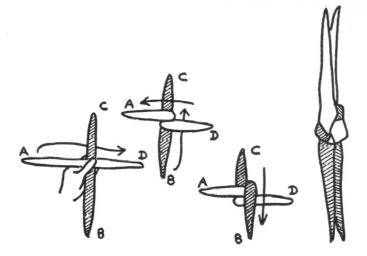

FIG. 23   FIG. 24   FIG. 25   FIG. 26

Knotty Problems

Another handkerchief stunt, topological in character, is to challenge anyone to grasp a handkerchief by opposite corners and tie a single knot in the center without letting go of either end.   This is accomplished by twisting the cloth rope fashion and placing it on a table.   The arms are then folded.   With the arms still crossed, bend forward and pick up an end of the handkerchief in each hand.   When the arms are unfolded, a knot automatically forms in the center of the cloth.   Topologically speaking, the arms, body, and handkerchief are a closed curve in the form of a "trefoil" knot.   Unfolding the arms merely transfers the knot from the arms to the cloth.

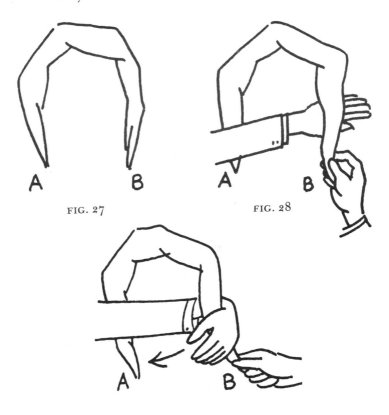

FIG. 27   FIG. 28

FIG. 29

An entertaining variation of this stunt can be performed
with a piece of rope or a man's scarf. It is placed on a table
as shown in Fig. 27. Grasp end *B* with your right hand and
ask the spectators to observe closely your method of tying a
knot. Slide your left hand beneath end *B*, palm down (Fig.
28), then rotate the hand back as in Fig. 29 to pick up end *A*.
When you separate your hands a knot will form in the scarf.
For some reason this move is extremely difficult to follow.
You can do the trick over and over again, but when others try
to duplicate it, no knot forms.

STRING AND ROPE

Innumerable tricks and stunts with string may be regarded as
topological. (The most complete compilation of string tricks
is Joseph Leeming's FUN WITH STRING, Frederick A. Stokes Co.,
1940.) Most of them employ a piece of cord with ends knotted
to form a simple closed curve. This loop can be twisted
around the fingers in certain ways, until it seems hopelessly
entangled, then a pull frees it easily. Or it can be looped
about a spectator's finger, through a buttonhole, or around the
head or foot, and pulled free in a manner that seems to violate
topological laws. It can be placed over a spectator's thumbs,
a finger ring threaded on it, and the ring removed without
taking the loops from the thumbs.

Garter Tricks

A category of topological string tricks is known as "garter
tricks," apparently because they were once performed with
garter belts in days when men wore high silk stockings. The
magician first arranges the cord to form a pattern on the top
of a table. A spectator tries to place his finger inside one of
the loops so that when the magician pulls the string away, the
cord will catch on the finger. There are many ingenious patterns that enable the magician to control the game, causing
the string to catch or pull free regardless of where the spectator
places his finger. Fig. 30 shows the simplest version. The
spectator has a choice of loop *A* or *B*, but regardless of which

he chooses, the performer can cause him to win or lose by the manner in which he gathers the ends. Arrows C and D indicate the two ways the ends can be brought together.

This form of the garter trick can also be performed with a man's leather trouser belt by doubling it, then twisting it into a spiral by inserting the forefinger into the loop and twisting the double belt with the finger and thumb. As it is being twisted, spectators try to keep their eyes on the loop. They are asked to place their finger into what they believe to be the loop, but the performer invariably pulls the belt free. As in the string version, the performer can make the belt catch or pull free as he wishes.

More complex versions of the garter game have been described in books on magic as well as in works on swindling. (See Leeming's book, cited above, p. 5; Hans Gross, CRIMINAL INVESTIGATION, English translation, London, 1924, p. 563; and Dr. L. Vosburgh Lyons' subtle variation in MAGIC AS A HOBBY, by Bruce Elliott, p. 70.)

The Giant's Garter

A curious trick, similar to the garter trick, makes use of a string twenty feet or more in length. The ends are tied together to form a closed curve. Someone is asked to place the string on the rug in as complex a pattern as he wishes (see Fig. 31), the only provision being that the string does not cross itself at any point. After the pattern is formed, newspapers are placed around the edges as shown in Fig. 32 so that only a rectangular interior section of the pattern is visible.

A spectator now places his finger on the pattern at any spot he wishes, holding his finger firmly on the carpet. The question is this. If one of the newspapers is removed and an outside part of the string pulled horizontally across the floor, will or will not the cord catch on the spectator's finger? The complexity of the pattern, together with the fact that its outside edges are concealed, makes it seem impossible to know which spots on the carpet are inside the closed curve of the string and which spots are outside. Nevertheless the performer is able to state correctly, each time the experiment is tried, whether the string will or will not catch.

Another presentation of the stunt makes use of a dozen or more dime-store hat pins. The performer quickly pushes them, apparently at random, into the visible portion of the pattern until the rectangular space is dotted with pins. When the string is pulled across the carpet, it pulls free of every hat pin. One hat pin, of a different color, may be so placed that when the string is pulled it will pull free of all pins except the odd-colored one on which it catches. Still another variation is to place the pins *inside* the closed curve. In this case, the pulled string forms a loop surrounding all the pins.

These stunts are made possible by a few simple rules. If two points on the pattern are both inside the curve of the string, an imaginary line connecting them will always cross an *even* number of strings. If two points are both outside, the same rule applies. But if one point is inside and one outside, a line connecting them will cross an *odd* number of strings.

As the newspapers are being placed, let your eye move into the pattern from the outside, as though working a maze, until you reach a space near the center that is easy to recall. For example, you may remember space A in Fig. 31. This is a space you know to be outside. After the papers are placed it is now a simple matter for you to determine whether any given spot is inside or outside. You have only to draw an imaginary line (it need not be straight, but of course a straight line is the simplest to visualize) from the spot in question to the spot you know to be outside, and note whether you cross an odd or even number of strings.

The working of all the variations should be clear. A dozen hat pins can be placed rapidly outside the closed curve. Simply place the first pin outside, then cross two strings and place another, cross two more, place another, and so on. If you desire a single pin to catch, cross one string from any of the other pins before you push it into the rug. Of course you can just as quickly place all the pins inside the closed curve.

If you wish to take a chance, keep your back turned until the pattern has been formed and the newspapers placed around it. In most cases you can still spot the outside areas by looking along the borders for adjacent strings that are slightly concave to each other. Such strings are likely to surround an outside area. Convex adjacent strings, on the other hand, are likely to enclose an interior space. However, these two rules are not infallible. The best plan, if you do not steal a glance at the pattern before the newspapers are down, is to place the pins without stating in advance what you intend to accomplish. Then when the string is pulled, it will either pull free of all the pins or it will enclose all of them—either effect being equally surprising.

A similar trick can be presented with pencil and paper. Let someone draw a complex closed curve on a sheet of paper (the lines of course must not cross), then fold back all four edges so that only a rectangular inner section can be seen (Fig. 33). Ask him to draw X's at half a dozen spots on the pattern. You take the pencil and quickly circle all the X's which are inside. The edges of the paper are unfolded and your selection of X's is found to be correct.

More String Tricks

Still another category of topological string stunts involves tying the wrists with a single piece of cord as shown in Fig. 34. The string can be manipulated in such a way that a simple knot, or a figure-eight knot, can be formed in the cord. A rubber band can be placed on the string, or removed, without cutting or untying the cord. If two people are tied in this manner, with the strings interlocked as shown in Fig. 35, it is possible to manipulate the string so that the couple can be separated. An amusing party diversion is to link everyone in the room into couples, offering a prize to the pair who can become unlinked first. Couples will undertake astonishing contortions in fruitless efforts to free themselves.

The solutions to the above problems all depend on the fact that the circuit formed by the string, arms, and body is not a true closed curve, but one separable at the wrists. The knot is formed by passing a loop of string under the band circling one of the wrists, giving it a twist, then bringing it back over the hand, under the band again, and over the hand once more.

The figure-eight knot is formed the same way, except the loop is given two twists. The rubber band is placed on the cord by putting it over the hand, slipping it under the cord so it encircles the arm above the wrist, then carrying it over the hand and on to the string. Reversing these moves will, of course, remove it. The joined persons are separated in a similar manner by passing the center of one cord under the string circling the other person's wrist, over his hand, then back under the string again.

A very old trick with three beads and two pieces of string, the principle of which has been applied to many other effects with ribbons and ropes, is known in the magic profession as "Grandmother's Necklace." The beads are first shown threaded on two pieces of cord. When a spectator pulls on the ends, the beads drop from the string into the magician's hands.

Fig. 36 is a cross-section showing how the beads are threaded. The two pieces of string appear to go through all three beads, but actually each piece is doubled back as shown. Two ends of the cord are crossed, as pictured in Fig. 37. When the ends are pulled (Fig. 38) the beads fall from the cord.

A number of methods used in cutting and restoring a rope have topological aspects, as well as many curious ways of causing knots to form or dissolve while both ends of a rope are apparently held at all times.

A knot known as the Chefalo Knot is typical of dozens of remarkable false knots which have been developed by magicians. It begins as a legitimate square knot (Fig. 39). Then one end is woven in and out as shown in Fig. 40. When the ends are pulled, the knot dissolves readily.

For an excellent collection of magic knots, see THE ASHLEY BOOK OF KNOTS, by Clifford W. Ashley, 1944. Milbourne Christopher, a magician who features a routine of rope knots in his night club act, provided much of the material for the magic section of this exhaustive work. Other unusual knot effects, as well as many other rope tricks with topological features, will be found in THE ENCYCLOPEDIA OF ROPE TRICKS, by Stewart James, revised 1945 edition.

CLOTHING

Three entertaining topological parlor stunts make use of a gentleman's vest. (Topologically speaking, a vest may be regarded as a bilateral surface with three unlinked edges, each a simple closed curve. Buttoned, it becomes a bilateral surface with four such edges.)

The Puzzling Loop

A man wearing a vest is asked to remove his coat. After a loop of string has been placed over his arm, he hooks his thumb into the lower vest pocket as shown in Fig. 41. Others now try to remove the string without dislodging his thumb. The secret is to push the loop through the arm hole of the vest, carry it over his head, out through the other arm hole, and over his other arm. The loop will then be circling his chest beneath the vest. Lower it until it clears the vest and allow it to drop to the floor.

Reversing the Vest

A man with a vest removes his coat, then clasps his hands in front of him. Can his vest be turned inside out without unclasping his hands? It is done by unbuttoning the vest, lifting it over his head so it hangs on his arms, turning it inside out through the arm holes, and returning it to its former position.

Surprisingly, the same feat is topologically possible without unbuttoning the vest, the only difficulty being that the buttoned vest is too tight to permit pulling it over the head. But it can be demonstrated easily by substituting a pull-over sweater without buttons, for the vest. The sweater is manipulated exactly like the vest. An easy way to perform the feat on yourself is to tie your wrists together with a piece of cord, leaving a foot or so between the wrists to give freedom of movement. You will find it a simple matter to pull the sweater over your head, turn it inside out through the sleeves, and return it again.

If a coat is worn over the vest, and the hands tied, it is still possible to reverse the vest. The coat is first removed over the head and allowed to hang on the arms. The vest is then turned inside out as described above, the arm holes passing over the coat. After the reversed vest is back in place, the coat goes over the head and back on the body again.

Removing the Vest

It is possible to remove a man's vest without removing his coat. The simplest method is as follows. First unbutton the vest. Now tuck the left side of his coat into the left arm hole of the vest from the outside. Work the arm hole over his left shoulder then down over his left arm. The hole will now be circling the coat in back of the left shoulder. Continue working the hole around the body, passing it over his right shoulder and arm, and finally bringing it free of the right side of his coat. In other words, the arm hole circles completely around the body.

The vest will now be hanging on his right shoulder, beneath the coat. Push the vest half-way down the right coat sleeve, then reach up the sleeve, grab the vest, and pull it out through the sleeve.

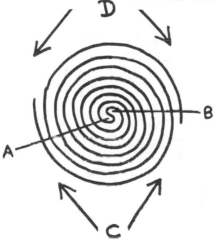

FIG. 30

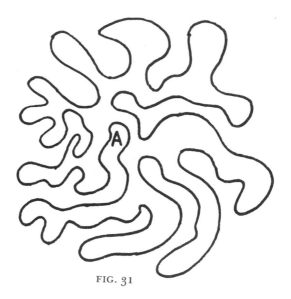

FIG. 31

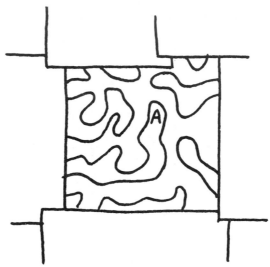

FIG. 32

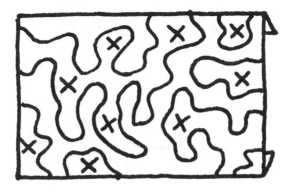

FIG. 33

FIG. 34

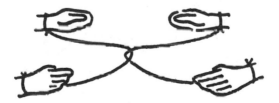

FIG. 35

FIG. 36

FIG. 37

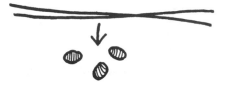

FIG. 38

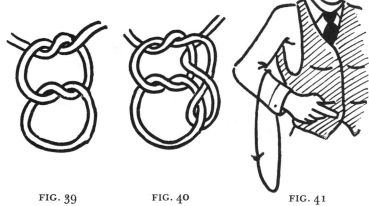

FIG. 39   FIG. 40   FIG. 41

RUBBER BANDS

We have already mentioned the trick of removing an elastic band from a string tied between the wrists. Here are two more rubber band tricks that are topological.

The Jumping Band

Place the band on the index finger (Fig. 42). Carry the other end around the middle finger (Fig. 43) and slip it back over the first finger once more (Fig. 44). Be sure that the band is looped around the fingers exactly as shown. Ask someone to hold the tip of your first finger.

As soon as he grasps the finger, bend your middle finger (Fig. 45). If the band has been properly placed, a portion of it will slip from the end of the middle finger. This causes the elastic to jump entirely free of the index finger and hang from the middle finger as shown in Fig. 45. It is difficult for others to duplicate this odd little feat. The stunt was originated by Frederick Furman, New York City amateur magician, who described it in *The Magical Bulletin*, Jan., 1921.

The Twisted Band

Another unusual stunt with an elastic band was contributed by Alex Elmsley to the January 8, 1955, issue of *Abracadabra*, a British magic journal. A large wide band is used. It is held as shown in Fig. 46. By sliding the right thumb and finger in the directions indicated by the arrows, the band is given two twists as shown in Fig. 47.

Ask someone to take the band from you by grasping it in exactly the same manner. In other words, his right thumb and finger take the top of the band from your right thumb and finger, and his left hand similarly takes the lower end of the band from your left hand. He will then be holding the twisted band exactly as you were holding it.

Challenge him to remove the twists from the band by changing the positions of his hands. He must not, of course, alter his grip on the two ends. Regardless of how he moves

FIG. 42   FIG. 43

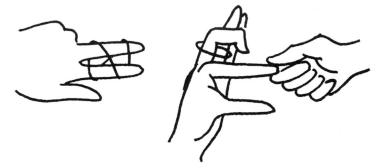

FIG. 44   FIG. 45

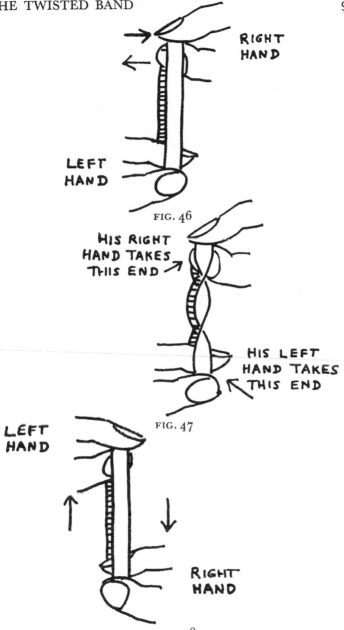

RIGHT HAND

LEFT HAND

FIG. 46

HIS RIGHT HAND TAKES THIS END →

HIS LEFT HAND TAKES THIS END

FIG. 47

LEFT HAND

RIGHT HAND

FIG. 48

his hands, he will discover that it is impossible to untwist the band.

You now carefully take the band back from him, holding it as you did originally. Then very slowly lower your right hand and raise the left as pictured in Fig. 48. When you do this, the twists mysteriously melt away.

Speaking topologically, what happens is this. The twisted band, together with your arms and body, form a structure of a type that permits an easy removal of the twists in the band. But when he takes the band from you there is a left-right reversal of part of this structure only. The result is a structure topologically different from the previous one.

Chapter Six

TRICKS WITH SPECIAL EQUIPMENT

Before 1900 magic was a highly esoteric and seldom practiced art. Since the turn of the century, however, and especially in the last two decades in the United States, it has experienced an extraordinary growth in the number of amateurs who have adopted magic as a hobby. In response to this increased demand for conjuring equipment, scores of magic supply houses have sprung into being. Catalogs of major American companies run to over 500 pages, and thousands of new tricks requiring specially made apparatus have been invented and placed on sale in the last decade alone.

Very few of these tricks, of course, are mathematical in character. But from time to time, mathematically-minded magicians devise apparatus that depend upon mathematical principles for their operation. It is from these items that I have chosen a few of the more interesting examples. In most cases the equipment can be constructed by the reader, but should he desire finished, factory-made material, I have indicated in the text where many of the items can be purchased.

Number Cards

I do not know when the first piece of mathematical magic equipment made its appearance, or what the item was, but certainly one of the most ancient tricks in this category is the set of cards for determining a person's age or the number he is thinking of.

The simplest version consists of a set of cards (usually six or more) each bearing a list of numbers. A person looks at every card, handing to the performer all cards that bear the number he has mentally selected. By glancing at these cards, the performer is able to name the number. It is obtained by adding together the lowest figure on each card. Since the numbers are usually arranged in order from low to high (to make it easy for the spectator to determine if his number is or is not on the card) these "key" numbers can be quickly spotted. The key numbers begin with 1, then proceed in a series obtained by doubling the previous number. Thus if six cards are used, the numbers are 1, 2, 4, 8, 16, and 32. The various combinations of cards will give totals from 1 to 63. In some versions, each card is a different color. This enables the performer, who has memorized the key numbers for each color, to stand across the room while the spectator is sorting the cards, and name the selected number without seeing the cards' faces.

Window Cards

A slightly more complex version, pictured in Fig. 49, makes use of the "window" device for obtaining the key numbers. After receiving all cards bearing the thought-of number, the magician puts them together, then places on top the "magic card" (at top of picture). Numbers showing through the holes are added to obtain the chosen number.

In principle, this set of cards is identical with the former set. The numbers, however, are not arranged in order, so the key numbers (i.e., the lowest number on each card) are in varying positions. The holes in the "magic card" correspond to these positions, and each card has a hole in every position except the spot where its own key number appears.

A still more complex form, using the window device, eliminates the necessity of adding key numbers. After the cards have been correctly assembled, only one number—the selected one—is visible through the windows. Many different ways to construct such cards are possible. Ball's MATHEMATICAL RECREATIONS AND ESSAYS and Kraitchik's MATHEMATICAL RECREATIONS contain descriptions of cards of this sort. Figs. 50 and 51 picture a set marketed several years ago in Winnipeg, Canada. Seven cards are used, numbered A to G. The

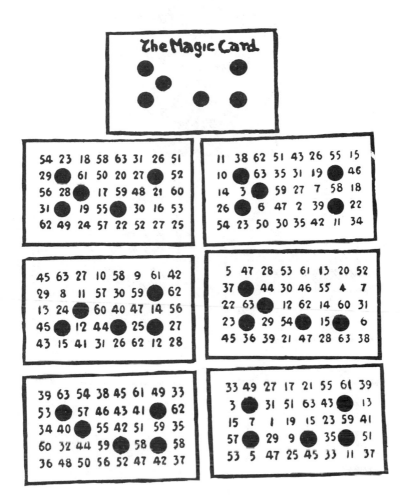

FIG. 49

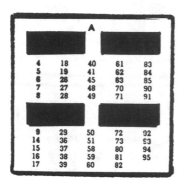

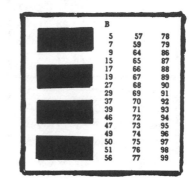

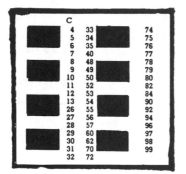

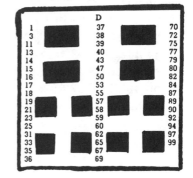

FIG. 50

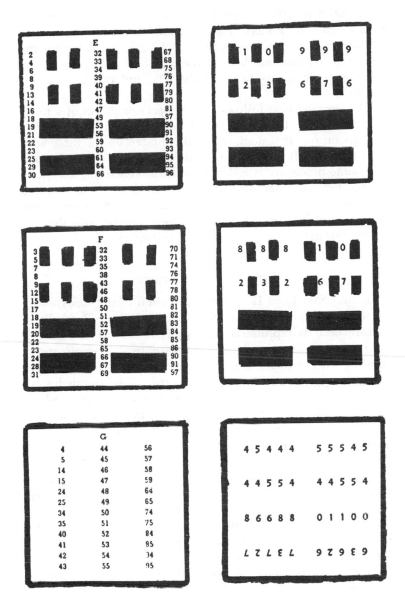

FIG. 51

backs of A,B,C,D are blank, but numbers appear on the reverse sides of E,F,G. Fig. 50 shows the first four cards, and above them the box containing the cards. Note that the box itself has four windows at the top. Fig. 51 shows the remaining three cards—faces on the left, and reverse sides on the right.

The working is as follows. The spectator thinks of a number under 100. He is handed card A and asked if his number appears on it. If it does, the card is placed on the table with A at the top. If not the card is turned so A is at the bottom. The same procedure is followed with the remaining six cards, taking them in order, and placing each on the preceding one after the spectator returns it. The cards are now turned over and inserted into the box. The chosen number will then be visible through one of the windows of the box.

Sam Loyd's Version

In 1924 Sam Loyd II patented a version of the cards which he sold under the name of "Sam Loyd's Telltale Puzzle." A picture of it is reproduced in SAM LOYD'S PICTURE PUZZLES, 1924. It consists of a large card with two windows and a rotating wheel in back. Six rectangles bearing numbers from 13 to 59 appear on the face of the card. The rectangles are numbered 1 to 6. To operate the device, you note which rectangles contain your age, then turn the wheel until the numbers of these rectangles appear in the upper window. A lower window, representing the page of a witch's book, then discloses your age.

The window principle can easily be applied to cards bearing names or pictures of objects instead of numbers. In 1937 Royal V. Heath created a set of six cards called "Think a Drink," which were manufactured by a New York premium firm. Properly assembled, the windows reveal the name of the mentally selected drink.

E. M. Skeehan, of Tulsa, Okla., many years ago applied the window principle to a set of cards that determine the day of the week for any given date over a period of 200 years. (Heath, it should be mentioned, in 1935 marketed a device for determining weekdays of dates from 1753 to 2140, but it made use of a slide rule rather than window cards.)

TAP TRICKS

Crazy Time

In Chapter Four a watch trick was explained in which the spectator thinks of an hour on the face, and the magician divines the hour by a process of tapping numbers until the spectator calls out "Stop." A more elaborate version of this trick, designed for presentation to a large audience, was placed on the magic market about fifteen years ago under the name of "Crazy Time." It is the invention of magician Tom Hamilton. Fig. 52 shows the front and back of the clock. The working is as follows:

The magician asks someone in the audience to select an hour in his mind and write it on a slip of paper without letting anyone see what he has written. A second spectator is asked to call aloud any number from 13 to 26 inclusive. The magician turns the board around so the back, bearing scrambled letters of the alphabet, faces the audience. With his wand he starts tapping the letters, apparently at random. The first spectator (who selected the hour) counts to himself with each tap, beginning with the number immediately above the hour he has written down. For example, if he has written "4," he counts "5" to himself on the first tap, then "6" on the second, and so on. When the count reaches the number called aloud by the second spectator, he says "Stop." The magician inserts his wand into the hole adjacent to the letter just tapped, and turns the board around. The wand is seen piercing the clock dial at the hour mentally selected by the first spectator.

Method: The magician subtracts 12 from the number called by the second spectator. Let us assume the number is 18. After 12 is subtracted, the remainder is 6. He makes the first five taps at random, but the *sixth* tap is made on the letter A. The succeeding taps are made on letters which spell the word "Ambidextrous," though of course the audience does not know

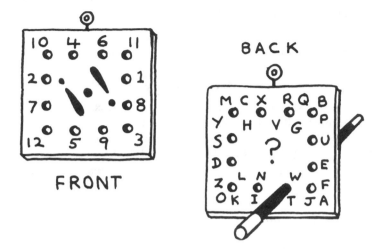

FIG. 52

this. He ceases tapping when the first spectator calls out "Stop," inserts his wand in the hole indicated, and when the clock is turned around, the wand is seen at the number thought of. A neatly painted board for use in this trick can be obtained from the Abbot Magic Co., Colon, Mich.

Heath's "Tappit"

An application of the tap principle to the *spelling* of numbers was made by Heath in 1925 when he marketed "Tappit." The trick employs six small tiles, each bearing a number, and each a different color. They are pictured in Fig. 53.

The tiles are placed on the table, numbered sides down. While the magician's back is turned a spectator looks at the number on one of the tiles, then shuffles them about. The performer turns around and begins tapping the tiles with a pencil. As he taps, the spectator spells the number silently to himself, calling out "Stop" when the spelling terminates.

FIG. 53

The tile on which the pencil is now resting is turned over. It proves to be the tile bearing the chosen number.

Method: The first six taps are made at random. The next six touch the tiles in the following order: 16, 13, 49, 85, 88, 77. The performer is able to tap them in this order because he has memorized the corresponding color sequence. The working depends, of course, on the fact that "sixteen" spells with seven letters, and each succeeding number has one additional letter in its spelling. A set of plastic tiles, with complete instructions, can be obtained from any magic dealer.

Tap-a-drink

Many other applications of the spelling principle have been made to magic tricks and novelty advertising cards. In 1940 I applied it to a give-away premium called the "Magic Tap-a-Drink Card." The front of this card is pictured in Fig. 54. The spectator thinks of one of the drinks. The card is turned over and the performer begins tapping the holes with his pencil. At each tap the spectator spells a letter to himself, calling "Stop" when the spelling is completed. The pencil is inserted in the last hole tapped. When the card is turned face up, the

FIG. 54

pencil is found to be in the hole by the chosen drink. The
first tap must be made on the top hole. Every other hole is
then tapped, proceeding in a clockwise direction.

Tap-an-Animal

A similar mind-reading trick which I contributed to *Children's Digest*, Dec., 1952, is shown in Fig. 55. A spectator
thinks of one of the animals pictured, then spells its name
silently as the magician taps the pictures. Taps begin on the
butterfly, follow the line upward to rhinoceros for the second
tap, then continue along the line to the other animals until

FIG. 55

the spectator calls out "Stop" at the termination of the spelling.

Numerous other applications of the principle have appeared. Walter Gibson has a tap trick with numbers in his *Magician's Manual*, employing cardboard polygons of different shapes. The "State Line" trick, marketed by Merv Taylor, is similar to "Tap-a-Drink" except that names of states are used instead of drinks. John Scarne created a premium card in 1950 titled "Think of a Number" in which only odd numbers were pictured on the card. This permits the magician to make every other tap at random while the spectator is counting to himself. (See SCARNE ON CARD TRICKS, Trick No. 97, for a card-tapping trick worked out on the same basis as his premium card.)

In all tap tricks of this sort a further mystification can be achieved by having the spectator spell his own name silently after he has finished the usual spelling or counting. To make it work you have merely to spell his name to yourself with random taps *before* you start tapping in the required pattern.

The Riddle Card

The coin-tapping trick explained in Chapter Four, in which the coins are arranged to form a figure 9, can also be adapted in many ways to premium cards. Fig. 56 shows one such card that I worked out many years ago. You choose the riddle you wish answered, tap up and around the circle as in the coin trick, then back around the circle until you end on one of the holes. The pencil point is pushed through the hole and the card turned over. A line on the back of the card leads from each hole to the proper answer.

DICE AND DOMINO TRICKS

Heath's "Di-ciphering"

In 1927 Royal V. Heath marketed an effect called "The Di-Ciphering Trick," based on a trick developed by Edmund Balducci. It consists of five cubes bearing a different three-figure number on each face—thirty numbers in all. The spec-

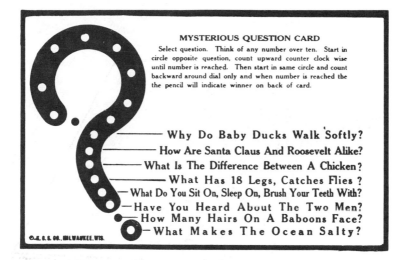

MYSTERIOUS QUESTION CARD

Select question. Think of any number over ten. Start in circle opposite question, count upward counter clock wise until number is reached. Then start in same circle and count backward around dial only and when number is reached the the pencil will indicate winner on back of card.

——— Why Do Baby Ducks Walk Softly?
——— How Are Santa Claus And Roosevelt Alike?
——— What Is The Difference Between A Chicken?
——— What Has 18 Legs, Catches Flies ?
—— What Do You Sit On, Sleep On, Brush Your Teeth With?
—— Have You Heard About The Two Men?
—— How Many Hairs On A Baboons Face?
—— What Makes The Ocean Salty?

C.J. E.S. CO., MILWAUKEE, WIS.

FIG. 56

tator rolls the cubes on the table and the magician quickly announces the total of the five uppermost numbers.

The trick is accomplished by adding the last digit in each number, then subtracting the total from fifty. The result is prefixed to the previous sum, which results in the grand total of all five numbers showing on the cubes. For example, we will assume that adding the last digits gives a total of 26. Subtracting 26 from 50 results in 24. The final answer, then, is 2426.

The five dice are numbered as follows: (1) 483, 285, 780, 186, 384, 681. (2) 642, 147, 840, 741, 543, 345. (3) 558, 855, 657, 459, 954, 756. (4) 168, 663, 960, 366, 564, 267. (5) 971, 377, 179, 872, 773, 278. It is a simple matter to make a set of the dice by penciling the numbers on the faces of five sugar cubes. A permanent set made of white plastic is obtainable in magic shops. For other tricks which can be performed with the dice, see MY BEST, edited by J. G. Thompson, Jr., 1945, p. 242f., and ANNEMANN'S PRACTICAL MENTAL EFFECTS, 1944, p. 59.

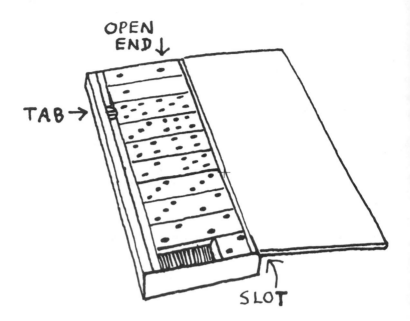

FIG. 57

Sure-Shot Dice Box

From time to time magic devices employing ordinary dice have been marketed, most of them combining the seven-principle (the fact that opposite sides of a die total seven) with other ideas. A good example is the dice box usually sold under the name of "Sure-shot dice box." (In *The Sphinx*, Sept., 1925, p. 218, the invention of the box is ascribed to Eli Hackman.) Shaped like a circular pill box, it permits dice to rattle noisily inside when the box is shaken, but is carefully designed to prevent the dice from actually turning over. In addition, the top and bottom of the box are alike, so that the box may be turned either way before it is opened.

Many tricks can be performed with this box, the best of which, in my opinion, was contributed by Stewart James to the August, 1949, issue of *The Linking Ring*. The magician first writes a prediction on a slate and puts the slate aside without showing its face. Three dice are placed in the box. A spectator shakes the box, opens it, and records the total showing on the three dice. This is repeated six more times, making seven totals in all.

At this point the magician gives the spectator a choice between two procedures. He may stop now and add the seven totals, or he may shake the dice two more times, making nine totals to be added. Whichever he decides, the grand total corresponds exactly to the number which the performer previously wrote on the slate.

The trick works as follows. When the box is first handed to the spectator, the faces of the three dice must total 5. He shakes the box, opens it, and records the total of 5. The magician picks up the dice, tosses them carelessly back into the box, closes it, and hands it out for the second shake which will of course result in a random number. After this number is recorded, the dice are left untouched, but the box is secretly inverted as it is handed out for the third shake. Thus the combined totals of the second and third shakes must equal 3 times 7, or 21. Once more the dice are picked up and tossed back into the box, and the same stratagem is followed for the next two shakes. The sixth and seventh shakes are similarly controlled. As a consequence, the grand total of all seven shakes must be 68 (3 times 21, plus the first total, 5). This is the number which the performer writes on the slate at the outset.

What, you may ask, does the magician do if the spectator chooses to make two additional shakes? He simply controls the shakes as before, resulting in an additional 21 which brings the grand total to 89. He then displays the slate upside-down to reveal 89 instead of 68—a brilliant touch that makes the effect extremely baffling.

Blyth's Domino Box

An interesting variation of a domino trick explained in Chapter Four was described in Will Blyth's EFFECTIVE CONJURING, 1928, and recently marketed under the name "Mental Domino Trick." Ten dominoes fit into a narrow plastic box (Fig. 57) which is open at the upper end. When the cover is closed, the lowest domino can be slid from the box to the right, then reinserted through the opening at the top. Along the left edge of the box is a tab that can be moved up and down. The performer sets the tab, then closes the box and requests the spectator to transfer one to ten dominoes from bottom to top. Let us suppose he transfers six. When the cover is opened, the tab will be found opposite a domino on which the spots add to six. The trick is immediately repeated without altering the positions of the dominoes.

To make the trick work, all the performer has to do is set the tab each time opposite the domino which has spots totaling ten. The box can be obtained from Louis Tannen, 120 W. 42nd St., New York, N.Y.

Blocks of India

A trick sold under the name of "The Blocks of India" is an adaptation of another domino trick previously described. The dominoes have no spots, but each block is divided into two colors. Many colors are used, so that no two blocks are alike. While the magician is out of the room, a spectator forms a chain of the domino-shaped blocks, matching color for color. In every case the magician is able to predict in advance the two colors which will be at the ends of the completed chain. As in the domino trick, the method consists in secretly removing one of the blocks before each performance of the trick. The two colors on this block will be the same as the colors at the ends of the chain. The trick is sold by the National Magic Co., Palmer House, Chicago.

Hummer Tricks

Many of Bob Hummer's mathematical tricks using cards and other common objects have been described in earlier chapters. In addition to these tricks, Hummer also has created a number of curious mathematical effects requiring special apparatus. Some of these tricks have been marketed; others exist only in the form of Hummer's home-made samples.

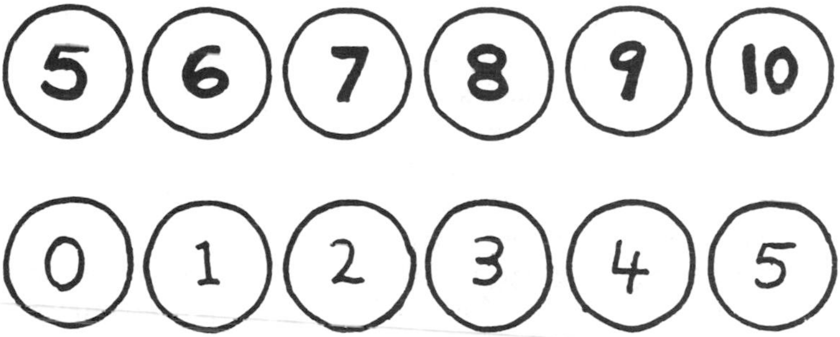

FIG. 58

One of Hummer's best-known items is his "Poker Chip Trick." Six poker chips are used, each bearing a number on both sides. In Fig. 58 the upper row shows the numbers on one side of the chips. Immediately below are shown the reverse sides. Note that the numbers in the upper row are in heavy, bold-face type. Those on the reverse side are in light-face type.

To perform the trick, the magician asks the spectator to shuffle the chips about in his hands, then place them on the table in two rows of three each. While the magician turns his back, the spectator turns over any three chips without telling the performer which ones they are. The magician gives instructions for turning over a few more chips. The spectator now touches any chip he wishes, turns it over, and covers it with a card (a playing card or business card—anything that will hide the chip). He repeats this with two other chips. There are now three exposed chips on the table and three covered. At this point the magician turns around, glances at the chips, and names the total of the three concealed numbers.

The working is as follows. Before turning his back the magician glances at the two rows of chips and remembers the positions of all chips that show *bold-face* numbers. After the spectator has reversed any three chips, the magician gives instructions for turning over a few more. For example, he may say, "Turn the second chip in the first row and the third chip in the bottom row." These chips that he orders reversed are the ones in the positions he is remembering (that is, the positions which were occupied by bold-face chips before he turned his back).

The spectator now reverses three chips, covering each with a card. The magician turns around and mentally performs the following calculations. He counts the number of chips which show bold-face figures (there will be none, one, two, or three) and multiplies this number by 10. To this product, 15 is added. From the sum is subtracted the total of the three exposed chips. The remainder is the total of the figures uppermost on the three covered chips.

For readers who might be interested, I list below three other Hummer creations, too complex to explain here, but currently available from magic dealers.

"Mother Goose Mystery." A printed booklet of standard Mother Goose rhymes used for performing two unusual mind-reading tricks. I worked out the booklet in 1941 on the basis of a suggestion by Hummer.

"It's Murder." A mathematical mind-reading trick employing a special board on which a circle bears the names of ten people. A five-pointed star is placed on the circle so that one of its points indicates the person to be "murdered." An ingenious method enables the performer to determine the name of the victim without knowledge of which point on the star has been chosen to represent the "murderer." A neatly-printed cardboard version of the trick can be obtained from the manufacturer, Frank Werner, 6948 Linden St., Houston, Texas.

"The Magic Carpet." A set of 26 cards, each bearing a letter of the alphabet, and a tiny cloth carpet are employed. The spectator conceals, beneath the carpet, cards bearing the letters of a girl's name, and the magician is able to determine the name without seeing the cards. The trick makes subtle use of a doubling principle. It can be purchased from the Sterling Magic Co., P.O. Box 191, Royal Oak, Mich.

To Bob Hummer and Royal V. Heath are due the honors of being the outstanding contemporary American inventors of mathematical magic tricks making use of special equipment. The term "mathemagic" was originated by Heath, and to my knowledge he is the first and only magician to present an entire show of original effects based on mathematics in their operation. It is regrettable that so many of the best mechanical creations of these two men have not been marketed, and are therefore unobtainable by others interested in this type of magic.

Chapter Seven

Geometrical Vanishes—Part I

In this and the next chapter we shall trace the development of a number of remarkable geometrical paradoxes, some old and some published here for the first time. All of them involve the cutting and rearranging of parts of a figure. After the rearrangement is completed, a portion of the original figure (either part of its area or one of a series of pictures drawn on the figure) has apparently vanished without trace! When the pieces are returned to their original form, the missing area or picture mysteriously appears once more. It is this curious vanishing and reappearing that justifies regarding these paradoxes as mathematical magic.

The Line Paradox

So far as I know, no one has recognized the fact that the various paradoxes to be discussed here all operate by a common principle. For lack of a better name, we shall call it the *Principle of Concealed Distribution*. The following elementary paradox (Fig. 59), which is very old, will make the principle clear.

You will note that ten vertical lines of equal length are so placed on the rectangle that if you follow the dotted diagonal from left to right, you find a progressive decrease in the length of the segments above the diagonal and a corresponding increase in the length of segments below. We shall now cut the rectangle along the diagonal and slide the lower part downward and leftward to the position shown in Fig. 60.

If you count the vertical lines in the above figure you will discover that there are now only nine. Which line vanished

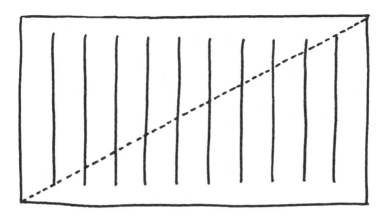

and where did it go? Slide the lower part back to its former position and the missing line returns. But which is the line that returned and where did it come from?

At first thought these are puzzling questions; but it takes only a superficial analysis to realize that no individual line vanishes. What happens is that eight of the ten lines are broken into two segments, then these sixteen segments are re-distributed to form nine lines, each a trifle longer than before. Because the increase in the length of each line is slight, it is not immediately noticeable. Actually, the total of all these small increases exactly equals the length of one of the original lines.

Perhaps the working of the paradox can be made even clearer if we consider five groups of marbles, each group containing four marbles. Let us shift one marble from the second group to the first, two marbles from the third group to the second, three from the fourth group to the third, and finally, all four marbles in the fifth group are moved to the fourth. Fig. 61 will make this clear.

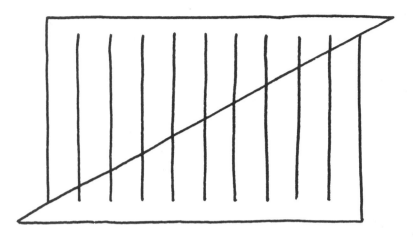

FIG. 60

FIG. 61

After the shifts have been made we will find that we have only four groups of marbles. It is obviously impossible to answer the question "Which group vanished?" because three of the groups were broken up and redistributed in such manner that each group gained an additional marble. This is of course exactly what occurs in the line paradox. As the shift is made along the diagonal, each line increases slightly in length as the segments of the broken lines are recombined.

Sam Loyd's Flag Puzzle

Sam Loyd based an interesting cutting puzzle on this paradox. The problem is to cut an American flag of fifteen stripes into the fewest number of pieces that can be fitted together to form a flag of thirteen stripes. Fig. 62, from Loyd's CYCLOPEDIA OF 5,000 PUZZLES, shows how the problem is solved by cutting the flag into two pieces.

FIG. 62

The Vanishing Face

Let us now consider other ways in which the line paradox can be elaborated to make the vanish and reappearance more interesting. Obviously, two-dimensional figures can be substituted for the lines. We can use pictures of pencils, cigars, bricks, tall silk hats, glasses of water, and other objects that can be drawn with vertical sides so the pictures will fit properly both before and after the shift is made. With a little artistic ingenuity, more complicated pictures can be used. Consider, for example, the vanishing face in Fig. 63.

When the lower strip is shifted to the left as indicated, all the hats remain but one face vanishes completely! It is useless to ask which face vanishes because after the shift is made, four of the faces have been broken into two parts and the parts redistributed so that each face gains a small amount—a longer nose here, more of a chin there, and so on. But the distribution is cleverly concealed, and of course the vanish of an entire face is much more startling than the vanish of a line.

"Get Off The Earth"

Sam Loyd must have been thinking along these lines when he invented and patented in 1896 his famous "Get off the Earth Puzzle." It was Loyd's greatest creation. More than ten million copies were reportedly sold here and abroad during Loyd's lifetime. During 1897 it was distributed by the Republican Party to promote McKinley's presidential campaign. At Robert Ripley's Odditorium, in Chicago's 1933 Century of Progress Exposition, the "talker" outside the building used a huge wooden reproduction of the puzzle as a device for attracting crowds. He would count the thirteen Chinese warriors with a pointer, spin the wheel rapidly, stop it at the desired position, then count the figures again to prove that one warrior had vanished. Loyd's original puzzle, as drawn by himself, is reproduced in Fig. 64. (For Loyd's account of the invention of this paradox, see "The Prince of Puzzle-makers," *Strand* magazine, Vol. 34, 1907, p. 771.)

What Sam Loyd did was simply bend the line paradox into a circular form and substitute figures of Chinese warriors for the lines. In the illustration ahead there are twelve figures. If the circle is cut, then rotated clockwise until the arrow points to N.E., the broken parts of the figures fit together once more to form thirteen warriors. When the arrow is returned to N.W., the extra figure vanishes.

You will note that when there are thirteen warriors showing, two of them are directly opposite each other at the lower left of the circle. These two figures correspond to the end lines of the line paradox. It is necessary that each of them lose a small portion of leg, and by bringing them together when the wheel is turned, the loss becomes less apparent. The wheel can be rotated further to produce fourteen, fifteen, etc., warriors, but as the number increases it becomes more obvious that each figure is being depleted to provide substance for the new figures.

The warriors are drawn with an ingenuity much greater than one might think at first inspection. For instance, in order to keep the figures in an upright position around the globe it is necessary at one spot for a left leg to become a right one, and at another spot for a right leg to become a left one.

During 1896, after the puzzle made its appearance, Loyd printed (in his Sunday puzzle column in the Brooklyn *Daily Eagle*) more than fifty letters from readers offering various explanations, many of them hilarious. Several readers were moved to versification. "I used to be a happy man . . ." begins a poem by Wallace Vincent, who goes on to say:

> But now I sit in solitude,
>   With aspect sad to see.
> My frame is thin, my eyes, deep sunk,
>   Burn fierce with lunacy.

> The only window in my room
>   Is crossed with bars of steel,
> And through an opening in the door
>   Attendants push my meal.

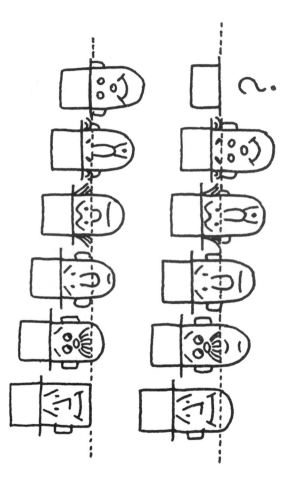

FIG. 63

FIG. 64

The Disappearing Chinaman

FIG. 64 a

From morn till night I crouch within
A corner of the room,
And gaze at something in my hand
With eagerness and gloom.

I push it up, I push it down,
I count them all once more,
And then I give a baffled shriek
And pound upon the floor.

I spare the reader the final stanzas of this sad ballad.

In 1909 Loyd designed a similar puzzle called "Teddy and the Lions" which pictured Theodore Roosevelt, seven lions, and seven African natives. The natives and lions alternated around the circle, one group spiraling inward, the other outward. Consequently, turning the wheel produced eight lions and six natives. The puzzle is now a rare collector's item. The only copy I have seen, owned by Dr. Vosburgh Lyons, was distributed to advertise the Eden Musee, a waxworks museum in Manhattan. The text on the back of the card reads in part:

"This mystery was designed by Sam Loyd, the man who originated Pigs in Clover, the Fifteen Block Puzzle, Parchesi, etc., but this is his greatest work. You see a BLACK man turn into a YELLOW lion right before your very eyes, but the more you study it the less you know! Professor Rogers says, 'It is an optical illusion, or is printed with phosphoric florescent ink, I don't know which!' Send the best answer you can, explaining the mysterious disappearance of the black man, to the Puzzle Editor of *The Globe*, as twenty-five free tickets to the Eden Musee will be distributed every week for the best answers. . . ."

DeLand's Paradox

In 1907 a Philadelphia engraver and amateur magician, Theodore L. DeLand, Jr., copyrighted another ingenious variation of the vertical line paradox. He printed it in several forms, one of which is shown in Fig. 65.

FIG. 65

The card is cut along horizontal line *AB* and vertical line *CD* to form three pieces. By exchanging the positions of the two lower rectangles, the same result is achieved as by sliding the lower part of the line paradox. One of the playing cards vanishes. An advantage of this construction is that the progressive, stairstep sequence of the figures is broken into sections so that the figures seem to be scattered over the card in a more random manner. To obscure the arrangement still further, DeLand added some extra cards which play no part in the paradox.

The Vanishing Rabbit

DeLand's paradox can obviously be elaborated by using more complicated pictures—faces, human figures, animals, and so on. Reproduced in Fig. 66 is a variation I designed for the Family Fun section of *Parents Magazine*, April, 1952. As you see, I merely turned the DeLand construction to a vertical position and substituted rabbits for playing cards. When rectangles A and B are switched, a rabbit disappears and in its place is an Easter egg. It would have been possible to make the bunny vanish completely, leaving an empty space, but an amusing Easter touch is added by having the tip of a nose and the end of a tail fit together to form the egg left by the Easter Bunny before he departed.

If instead of shifting A and B, the right half is cut along the dotted lines and the two pieces interchanged, the number of rabbits will increase to twelve. One rabbit loses his ears, however, and other grotesque results occur.

Stover's Variations

It remained for Mel Stover, of Winnipeg, Canada, to add a final touch to the DeLand idea. In 1951 Stover designed a card in which two sets of pictures are interwoven (Fig. 67) in the manner of Sam Loyd's "Teddy and the Lions" puzzle. One set consists of glasses of beer and the other set of men's faces. When the shift is made, a face vanishes and an extra glass of beer appears—presumably one man has turned into a

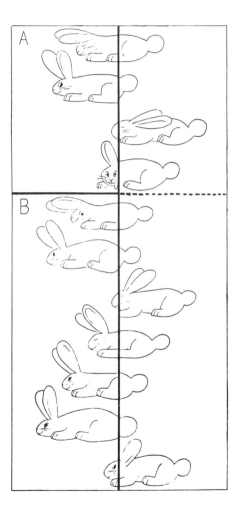

FIG. 66

glass of beer. The use of interwoven sets of pictures opens up many magical possibilities. Stover designed, for example, another card that pictures a number of pencils, some red and some blue (Fig. 68). Shifting two parts of the picture causes a blue pencil to vanish and a red pencil to appear, giving the impression that a pencil was made to change color.

As Mr. Stover points out, this paradox can be constructed in three dimensional form, using real pencils mounted on panels of wood. In fact, all the paradoxes discussed in this chapter and the next can be designed in 3-D forms, though in most cases nothing is gained by adding the new dimension.

In all forms of the DeLand paradox the number of units in the two pieces to be switched must be prime to each other— that is, they must have no common denominator other than 1. In the rabbit paradox, for example, part A contains 4 units and part B contains 7. If the total number of units is prime, as 11 is in this case, then the dividing line between A and B can be made wherever you please, because no two numbers the sum of which is a prime can have a common denominator. In all forms of the paradox the pictures can be jogged in such manner that the vanish occurs at any desired spot in the row.

The DeLand principle applies easily to Sam Loyd's circular form. Instead of having the Chinese warriors spiral gradually inward, the figures can be broken into smaller groups of stair-steps as in the DeLand form. In this case, the circle must be turned the distance of several units instead of one unit. The only advantage is that the pictures seem to be placed at random and the principle of the paradox is thereby better camouflaged.

Three-dimensional forms of the circle can be made, of course, simply by placing the pictures around the outside of a cylinder, cone, or sphere, which is cut so that one half can be rotated relative to the other.

FIG. 67

© MEL STOVER, 1956

© MEL STOVER, 1956

FIG. 68

Chapter Eight

Geometrical Vanishes—Part II

THE CHECKERBOARD PARADOX

Closely related to the picture paradoxes discussed in the last chapter is another class of paradoxes in which the Principle of Concealed Distribution is responsible for a mysterious loss or gain of area. One of the oldest and simplest examples is reproduced in Fig. 69.

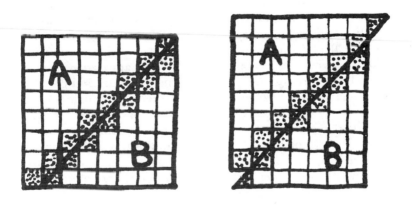

FIG. 69

The checkerboard on the left is cut along the diagonal line. Part B is then shifted downward as shown on the right. If the

projecting triangle at the upper right corner is snipped off and
fitted into the triangular space at the lower left, a rectangle of
7 by 9 units will be formed. The original square had an area
of 64 square units. Now we have an area of 63. What happened to the missing square?

The answer lies in the fact that the diagonal line passes
slightly below the lower left corner of the square at the upper
right corner of the checkerboard. This gives the snipped-off
triangle an altitude of 1 1/7 rather than 1, and gives the entire
rectangle a height of 9 1/7 units. The addition of 1/7 of a
unit to the height is not noticeable, but when it is taken into
account, the rectangle will have the expected area of 64 square
units. The paradox is even more puzzling to the uninitiated
if the smaller squares are not ruled on the figure. When the
square units are shown, a close inspection will reveal their inaccurate fitting along the diagonal cut.

The relation between this paradox and the vertical line
paradox of the last chapter will become clear when we examine
the smaller squares that are cut by the diagonal line. As we
pass up the line we find that portions of the cut squares
(shaded in the illustration) become progressively smaller above
the line and progressively larger beneath it. There are fifteen
of these shaded squares on the checkerboard, but only fourteen
after the rectangle has been formed. The apparent vanish of
a shaded square is simply another form of the DeLand paradox
discussed earlier. When we snip off the tiny triangle and replace it, we are actually cutting Part A of the checkerboard
into two pieces which are then switched in their positions
along the diagonal line. All the skullduggery is confined to
the small squares along the diagonal cut. The other squares
play no part whatever in the paradox. They are merely padding. But by adding them we change the character of the
paradox. Instead of the vanish of a small square in a series
of squares drawn on paper (or any more complex figure such
as a playing card, face, etc., which we can draw within each
small square), we have an apparent change of area in a larger
geometrical figure.

Hooper's Paradox

A similar paradox of area, in which the similarity to De-Land's principle is even more obvious, is found in William Hooper's RATIONAL RECREATIONS, 1794 edition, Vol. 4, p. 286.

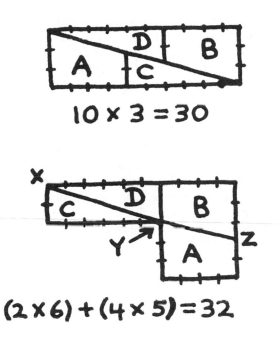

FIG. 70

By exchanging the positions of A and C as shown in Fig. 70, a rectangle of 30 square units is transformed into two smaller rectangles with a combined area of 32 square units—a gain of two square units. As before, only the squares along the diagonal are involved in the change. The rest is excess padding.

There are two basically different ways in which the pieces of Hooper's paradox can be cut. If we construct the rectangle at the top first and rule the diagonal accurately from corner to corner, then the two smaller rectangles in the figure on the right will each be 1/5 of a unit shorter than their apparent heights. On the other hand, if we construct the second figure first, drawing both rectangles accurately and ruling a straight line from X to Y and from Y to Z, then line XZ will not be perfectly straight. It will form an angle, with Y as the vertex, but so obtuse an angle that it appears to be a straight line. As a result, when the first figure is formed there will be a slight overlapping of the pieces along the diagonal. The previous paradox of the checkerboard, as well as most of the paradoxes to be discussed in this chapter, can likewise be constructed in two such variant forms. In one form there is a tiny loss or gain in the entire height (or width) of the figures. In the other form, the loss or gain occurs along the diagonal—either as overlapping, as in this case, or as an open space as we shall see presently.

Hooper's paradox can be given an infinite number of forms by varying the proportions of the figures and the degree of slope of the diagonal. It can be constructed so that the loss and gain is 1 square unit, 2, 3, 4, 5, and on up to infinity. The higher you go, of course, the easier it is to see how the missing square units are distributed—unless the rectangles are extremely large relative to the number of units caused to vanish.

Square Variation

An elegant variation of this paradox makes use of two rectangles of such proportion that they fit side by side to make a perfect 8 by 8 checkerboard. When the pieces are rearranged to make the larger rectangle, there is an apparent gain in area of one square unit. (See Fig. 71.)

If the square is constructed accurately, then the large rectangle does not have an accurate diagonal. There will be a rhomboid space along the diagonal, but so elongated that it is not discernible. On the other hand, if the large rectangle is drawn with an accurate diagonal, then the square's upper rectangle will be a trifle higher than it should be and the lower rectangle a trifle wider. The inaccurate fitting caused by the

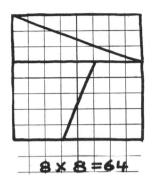

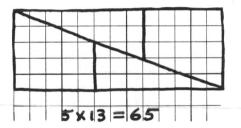

FIG. 71

second method of cutting is more noticeable than the inaccuracy along the diagonal in the first method, hence the first is preferable. As in previous examples, we can draw circles, faces, or other figures inside the squares cut by the diagonal lines, and these figures will gain or lose one of their members as we shift the positions of the pieces.

W. W. Rouse Ball, in MATHEMATICAL RECREATIONS AND ESSAYS, gives 1868 as the earliest date he could find for the published appearance of this paradox. The elder Sam Loyd, discussing the paradox on p. 288 of his CYCLOPEDIA OF 5,000 PUZZLES, states that he presented it to the American Chess Congress in 1858; and in his column in the Brooklyn *Daily Eagle* he once referred to it as "my old dissected checkerboard problem out of which sprang a close relative of the missing Chinaman." When you understand the paradox, he wrote, "you will know something about the Chinese methods of getting off the earth." It is hard to tell from these words whether Loyd claimed to have originated the paradox or merely to have been the first to introduce it to the public.

Sam Loyd's son (who adopted his father's name and continued his father's puzzle columns) was the first to discover that the four pieces could be put together in such a way that the area is reduced to 63 squares. Fig. 72 shows how this is done.

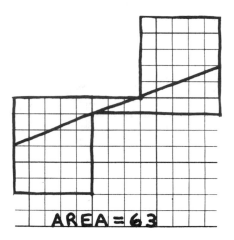

FIG. 72

Fibonacci Series

The lengths of various line segments involved in the four pieces fall into a Fibonacci series—that is, a series of numbers, each of which is the sum of the two preceding numbers. The series involved here is as follows:

1–1–2–3–5–8–13–21–34–etc.

The rearrangement of the pieces to form the rectangle illustrates one of the properties of this Fibonacci series, namely that if any number in the series is squared, it equals the product of the two numbers on either side of it, plus or minus 1. In this case the square has a side of 8 and an area of 64. In the Fibonacci series 8 is between 5 and 13. Since 5 and 13 automatically become the sides of the rectangle, the rectangle must have an area of 65, a gain of one unit.

Owing to this property of the series, we can construct the square with a side represented by any number in the series above 1, then cut it according to the two preceding numbers in the series. If, for example, we choose a 13-sided square, we then divide three of its sides into segments of 5 and 8, and

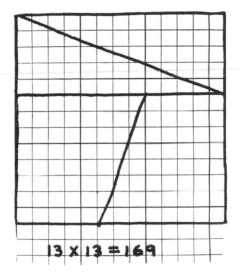

FIG. 73

rule the cutting lines as shown in Fig. 73. This square has an area of 169. The rectangle formed from the same pieces will have sides of 21 and 8, or an area of 168. Due to an overlapping along the diagonal of the rectangle, a unit square is lost rather than gained.

A loss of one square unit also occurs if we choose 5 for the side of the square. This leads us to a most curious rule. Alternate numbers in the Fibonacci series, if used for the square's side, produce a *space* along the diagonal of the rectangle and an apparent *gain* of one square unit. The other alternate numbers, used for the side of the square, cause an *overlapping* along the rectangle and a *loss* of one square unit. The higher we go in the series the less noticeable becomes the space or overlapping. Correspondingly, the lower we go, the more noticeable. We can even construct the paradox from a square having a side of only two units—but in this case the 3 by 1 rectangle requires such an obvious overlapping that the effect of the paradox is completely lost.

The earliest attempt to generalize the square-rectangle paradox by relating it to this Fibonacci series apparently was made by V. Schlegel in *Zeitschrift für Mathematik und Physik*, Vol. 24 (1879), p. 123. E. B. Escott published a similar analysis in *Open Court*, Vol. 21 (1907), p. 502, and described a slightly different manner of cutting the square. Lewis Carroll was interested in the paradox and left some incomplete notes giving formulas for finding other dimensions for the pieces involved. (See Warren Weaver's article, "Lewis Carroll and a Geometrical Paradox," in *American Mathematical Monthly*, Vol. 45 (1938), p. 234.)

An infinite number of other variations result if we base the paradox on other Fibonacci series. For example, a square based on the series 2, 4, 6, 10, 16, 26, etc., will give losses and gains of 4 square units. We can determine the loss and gain easily by finding the difference between the square of any number in the series and the product of its two adjacent numbers. The series 3, 4, 7, 11, 18, etc., produces gains and losses of 5 square units. T. de Moulidars, in his GRANDE ENCYCLOPEDIE DES JEUX, Paris, 1888, p. 459, pictures a square based on the series 1, 4, 5, 9, 14, etc. The square has a side of 9 and when

formed into a rectangle loses 11 square units. The series
2, 5, 7, 12, 19, etc., also produces losses and gains of 11. In
both these cases, however, the overlapping (or space) along the
diagonal of the rectangle is large enough to be noticeable.

If we call any three consecutive numbers in a Fibonacci
series A, B, and C, and let X stand for the loss or gain of area,
then the following two formulas obtain:

$$A + B = C$$
$$B^2 = AC \pm X$$

We can substitute for X whatever loss or gain we desire, and
for B whatever length we wish for the side of a square. It is
then possible to form quadratic equations which will give us
the other two numbers in the Fibonacci series, though of
course they may not be rational numbers. It is impossible,
for example, to produce losses or gains of either two or three
square units by dividing a square into pieces of rational
lengths. Irrational divisions will, of course, achieve these re-
sults. Thus the Fibonacci series $\sqrt{2}$, $2\sqrt{2}$, $3\sqrt{2}$, $5\sqrt{2}$ will
give a loss or gain of two, and the series $\sqrt{3}$, $2\sqrt{3}$, $3\sqrt{3}$, $5\sqrt{3}$
will give a loss or gain of three.

Langman's Version

There are many other ways in which rectangles can be cut
into a small number of pieces and the pieces rearranged to
form a rectangle of larger or smaller area. Fig. 74 pictures a
paradox developed by Dr. Harry Langman of New York City.

Langman's rectangle is also based on a Fibonacci series.
Like the square just discussed, a choice of alternate terms in
the series for the first rectangle's width (in this case 13) will
produce an increase of one square unit in the area of the
second rectangle. If one of the remaining alternate terms in
the Fibonacci series is used for the first rectangle's width, there
will be a loss of one square unit in the second rectangle. The
losses and gains are accounted for by a slight space or overlapping along the diagonal cut of the second rectangle. Another
version of Langman's rectangle, shown in Fig. 75, provides an
increase of two squares when the second rectangle is formed.

If we take the shaded portion of this rectangle and place

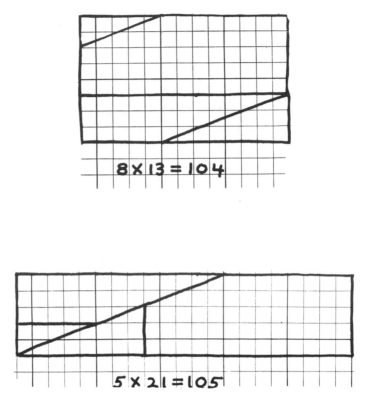

FIG. 74

it on top of the unshaded portion, the two diagonal cuts will form one long diagonal. Switching the positions of parts A and B will now form the second rectangle, of larger area. Thus we see that Langman's paradox is simply another form of the Hooper paradox discussed earlier.

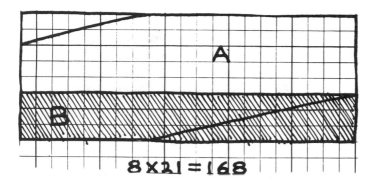

FIG. 75

Curry's Paradox

Let us now turn our attention to a simple form of Hooper's paradox. In the figure at the top (Fig. 76), an exchange in the positions of triangles B and C will cause an apparent loss of one square unit in the total area of the figure.

You will perceive that a change also occurs in the shaded areas. We have 15 shaded squares on the left, 16 on the right. By filling in these shaded areas with two oddly-shaped pieces, we arrive at a startling new manner of presenting the paradox. We have a rectangle that can be cut into five pieces which may then be rearranged to form a rectangle of identical size, but with a hole of one square unit within the figure! (Fig. 77.)

The inventor of this delightful paradox is Paul Curry, a New York City amateur magician. In 1953 he conceived the brilliant notion of cutting and rearranging the parts of a figure to form an identical figure with a hole inside its perimeter. In the above version of Curry's paradox, if point X is located exactly 5 units from the side and 3 units from the base, then

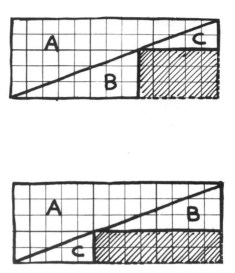

the diagonal line will not be perfectly straight, though the deviation will be so slight as to be almost undetectable. When triangles B and C are switched, there will be a slight overlapping along the diagonal in the second figure.

On the other hand, if the diagonal in the first figure is ruled accurately from corner to corner, then line XW will be a trifle longer than 3 units. As a consequence, the second rectangle will be slightly higher than it appears to be. In the first case, we may regard the missing square unit as spread from corner to corner, forming the overlap along the diagonal. In the second case, the missing square is distributed along the width of the rectangle. As we have seen earlier, all paradoxes of this type are subject to these two variant forms of construction.

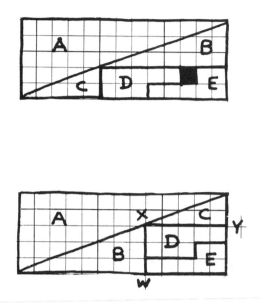

FIG. 77

In both forms the discrepancies in the figure are so minute that they are virtually invisible.

The most elegant forms of the Curry paradox are squares that remain square after the parts are reformed to show the hole. Curry worked out numerous variations, but was unable to reduce the parts to fewer than five pieces and still produce a hole that does not touch the border of the square. Curry squares have endless variations, with holes of any desired number of square units. A few of the more interesting forms are reproduced in Fig. 78.

Dr. Alan Barnert, New York City ophthalmologist, has called my attention to a simple formula which relates the size of the

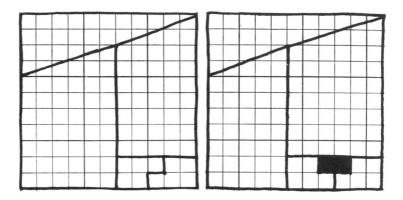

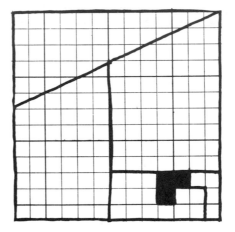

FIG. 78

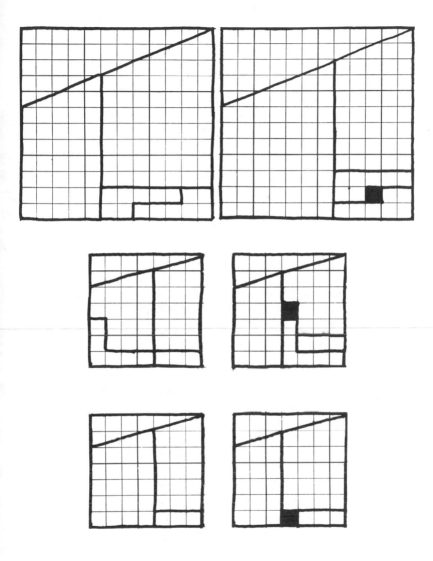

FIG. 78 CONTINUED

hole to the relative proportions of the three larger pieces. The three lengths involved are labeled A, B, and C in Fig. 79.

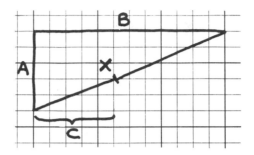

FIG. 79

The difference between the product of A and C, and the nearest multiple of B will give the number of square units in the hole. Thus in the above example, the product of A and C is 25. The multiple of B which is the closest to 25 is 24, therefore the hole will be one unit square. This rule holds regardless of whether the diagonal is ruled accurately as a straight line, or whether point X in the above figure is allowed to fall exactly on an intersection of the lattice. No paradox results if the diagonal is perfectly straight and point X is exactly on a lattice intersection. In such cases the formula gives zero as the size of the hole, meaning of course that there will be no hole at all.

Curry squares are best presented, in my opinion, by ruling the diagonal accurately so that the loss and gain is caused by a slight alteration in the square's height. If no lattice lines are ruled on the figure, this alteration will not be detectable. Small pictures may be drawn at each spot where a small square would be if the figure were crossed with lattice lines. One of these pictures will of course disappear. The hole suggests that the picture vanishes at the spot where the hole appears, though actually the vanish occurs, as we have seen, along the diagonal line.

Another entertaining presentation is to cut the pieces from wood, plastic, or linoleum—cutting them so the square will fit snugly into a box. Plug the hole with a small square piece. When you show the paradox, dump the pieces on a table, then replace them on their reverse sides. They are replaced in such manner that after the pieces are snugly fitted into the box, no spot remains for the small square piece! Most people find this extremely puzzling.

Royal V. Heath had a square made from polished metal pieces that fit into a small plastic case. Between the square and one side of the case is space for a plastic ruler to stand on edge. The ruler is the same length as a side of the square. It is removed first and used for proving that the case actually is square. The pieces are then removed and replaced in the case so that the hole appears. Because the ruler is no longer in the case, the case becomes longer by the thickness of the ruler, thus permitting the metal pieces to fit as snugly as before. No lattice markings appear on the pieces.

I should think it possible to bevel the edges of pieces in such a way that they overlap slightly when the pieces are in one formation, but do not overlap in the other formation. In this way the outside dimensions of the square could be kept exactly the same in both formations.

Curry Triangles

My own small contribution to this growing number of paradoxes is the discovery of simple triangular forms. Referring back to the first example of the Curry paradox, Fig. 77, you will note that the large triangle A remains in a fixed position while the other pieces are shifted. Since this triangle plays no essential role in the paradox, we can discard it entirely, leaving a right triangle cut into four pieces. The four pieces can then be rearranged (Fig. 80) to form an apparently identical right triangle with a hole.

By placing two of these right triangles side by side, a variety

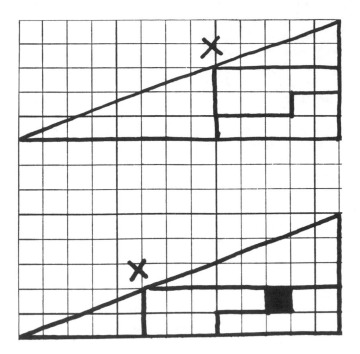

FIG. 80

of interesting isosceles forms can be worked out similar to that shown in Fig. 81.

Like the paradoxes previously discussed, these triangles can be constructed in two different ways. We can give them perfectly straight sides, in which cases points X will not fall precisely on intersections of lattice lines, or we can locate points X exactly on the intersections, in which case the sides will be slightly convex or concave. The latter method of cutting seems the more deceptive. This paradox is particularly surprising if lattice lines are ruled across the pieces, because they call attention to the accuracy with which the various pieces have been constructed.

Isosceles triangles may be given a variety of forms to show

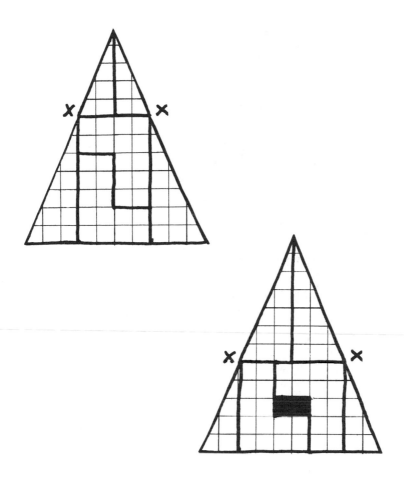

FIG. 81

gains and losses of any desired even number of square units. Reproduced in Fig. 82 are some representative samples.

If two isosceles triangles, of any of the types shown, are placed base to base, numerous variations of diamond shapes may be worked out, but they add nothing essentially new to the paradox.

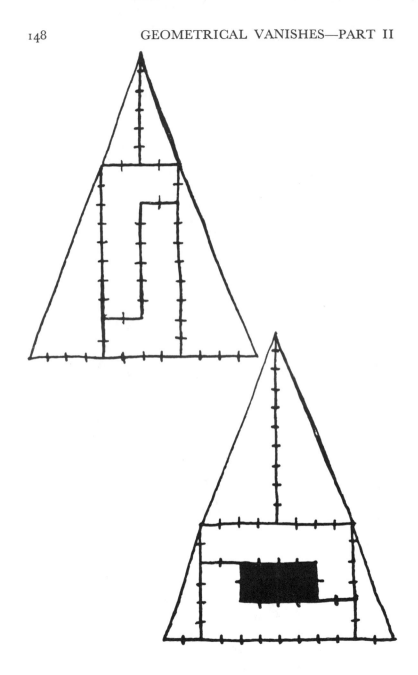

FIG. 82

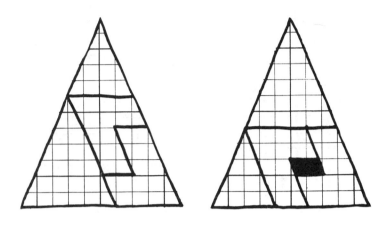

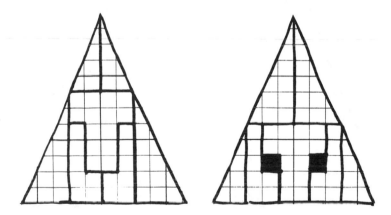

FIG. 82

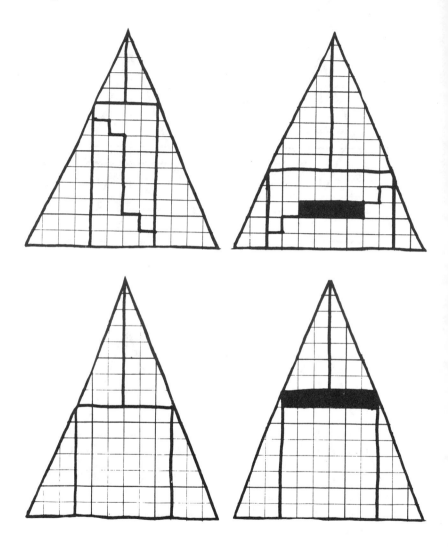

FIG. 82 CONTINUED

Four-piece Squares

All the forms of changing area paradoxes so far discussed
have been closely related in their mode of construction and
operation.   There are, however, other forms of quite different
construction.   For example, a square may be cut into four
pieces of identical size and shape (Fig. 83).   When the four
pieces are rearranged as shown in Fig. 84, they form a square
of apparently the same size, but with a hole of four square
units in the center.

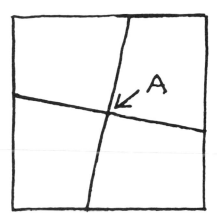

FIG. 83

A rectangle of any proportion can be cut in a similar manner.   Curiously, point *A* where the two perpendicular cuts
intersect can be at any spot on the rectangle.   In all cases a
space will appear when the pieces are rearranged, the size of
the space varying with the angle of the cuts to the sides.   The
area of the hole is of course distributed around the entire
perimeter of the rectangle.   The paradox has an attractive
simplicity, but suffers from the fact that only a casual

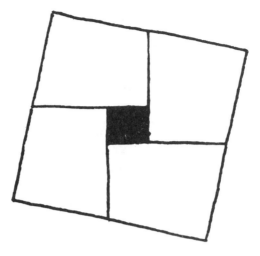

FIG. 84

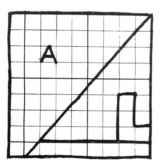 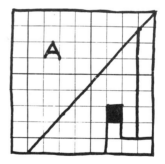

FIG. 85

inspection is needed to see that the sides of the second rectangle must of necessity be a trifle longer than the sides of the first.

A more perplexing manner of cutting a square into four parts to produce an interior hole is shown in Fig. 85. It is based on the checkerboard paradox with which this chapter opened. Two of the pieces must be turned over to form the second square. Note also that by eliminating part A we have a right triangle of three pieces in which an interior hole can be produced.

Three-piece Squares

Are there methods of cutting a square into three parts which can be fitted together to produce an inside hole? The answer is yes. One neat solution, devised by Paul Curry, is an adaptation of the DeLand paradox discussed in the previous chapter. Instead of jogging the pictures and making a straight horizontal cut, we place the picture units on a straight line and jog the cut. When this is done, a surprising result occurs. Not only does a picture vanish, but a hole appears at the spot where a picture is missing.

Two-piece Squares

Can it be done with two pieces? I do not believe it is possible to produce an interior hole in a square by any method that imperceptibly increases the height or width of the square. Paul Curry has shown, however, that it can be done by applying the principle explained above to Loyd's vanishing Chinese warrior paradox. Instead of spiraling or jogging the pictures, the pictures are placed in an even circle and the cutting is either spiraled or jogged like a series of different-sized cogs on a wheel. When the wheel is rotated, a picture vanishes and a hole appears. The parts fit snugly only when the hole is visible. In the other position there will be small spaces at each cog if the cut is jogged, or a space all the way around if the cut is a spiral.

If the rectangle to be cut is not a square, then it *is* possible to divide it into two pieces and form an inside hole by a subtle alteration of the outside dimensions. Fig. 86 shows an example I worked out in 1954. The two pieces are of identical size and shape. A simple way to demonstrate the paradox is to cut the pieces from cardboard, place them on a larger sheet of paper (in the form of a rectangle without the hole), then draw an outline around the perimeter. When the pieces are fitted together the other way, it can be seen that they still fit the outline, although a hole has appeared in the center of the rectangle. A third piece, in the form of a strip, can of course be placed along one side of this rectangle to make it a square, thus providing another method of cutting a square into three pieces that form an interior hole.

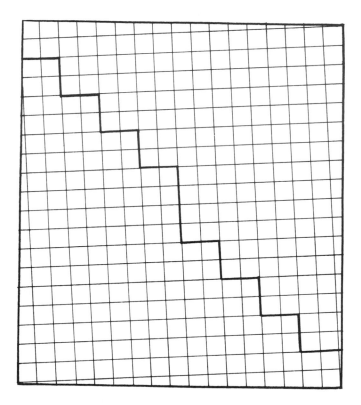

FIG. 86

Curved and 3-D Forms

It should be clear from these examples that the field of changing area paradoxes has only begun to be explored. Are there curved forms, such as circles or ellipses, which can be cut and rearranged to produce interior spaces without noticeable distortion of the figure? Are there three-dimensional forms that are unique in three dimensions—that is, not merely extensions of two-dimensional forms? Obviously, any of the figures discussed in this chapter can be given an added dimension merely by cutting them out of thick boards so that the hole extends the length of the third dimension. But are there simple ways of dissecting, say, a cube or pyramid so that when the parts are reassembled a sizable space forms in the interior?

With no limitations on the number of pieces, such solid forms are easily constructed. Curry's principle is clearly applicable to a cube in which an inside space appears, but to determine what the smallest number of parts can be is a more difficult matter. Such a cube can certainly be constructed with six pieces; other plans of dissection might reduce this number. A delightful presentation of such a cube would be to take it from a box that it completely fills, take it apart, find a marble in its interior, then put it together again to form a solid cube that fits snugly back into the box! There must be many forms, plane as well as solid, that have simplicity and elegance, and which future explorers in this curious realm will have the pleasure of discovering.

Chapter Nine

Magic with Pure Numbers

In this chapter we shall consider tricks in which only numbers are involved, aside of course from pencil and paper or a blackboard on which calculations may be made. Tricks of this sort fall in three general categories—lightning calculations, predictions, and mind-reading effects.

There is a considerable literature dealing with the first of these categories. Feats of mental calculation are, however, almost always presented as demonstrations of skill rather than as feats of magic. We shall do no more than glance at the four lightning calculation effects which have caught the fancy of conjurors. These are: (1) Naming the day of the week of any date called for (discussed briefly under calendar tricks in Chapter Four). (2) The chess knight's tour. (3) Constructing a magic square based on a total called by the audience. (4) The rapid calculation of cube roots.

The chess knight's tour has too often been discussed in the literature of recreational mathematics to warrant explaining here. Harry Kellar, a famous American magician who flourished around the turn of the century, used to feature the trick in his stage appearances (along with a demonstration of cube root extraction), but few magicians perform it today. Magic squares likewise arouse little interest among modern audiences. If the reader cares to learn a simple method of constructing a four-by-four magic square to fit a called-for total, he will find the stunt explained in Ted Annemann's BOOK WITHOUT A NAME, 1931.

Rapid Cube Root Extraction

The cube root demonstration begins by asking members of the audience to select any number from 1 through 100, cube it, then call out the result. The performer instantly gives the cube root of each number called. To do the trick it is necessary first to memorize the cubes of numbers from 1 through 10.

$$
\begin{array}{rcl}
1 & — & 1 \\
2 & — & 8 \\
3 & — & 27 \\
4 & — & 64 \\
5 & — & 125 \\
6 & — & 216 \\
7 & — & 343 \\
8 & — & 512 \\
9 & — & 729 \\
10 & — & 1000
\end{array}
$$

An inspection of this table reveals that each cube ends in a different digit. The digit corresponds to the cube root in all cases except 2 and 3, and 7 and 8. In these four cases the final digit of the cube is the difference between the cube root and 10.

To see how this information is used by the lightning calculator, let us suppose that a spectator calls out the cube 250047. The last number is a 7 which tells the performer immediately that the last number of the cube root must be 3. The first number of the cube root is determined as follows. Discard the last three figures of the cube (regardless of the size of the number) and consider the remaining figures—in this example they are 250. In the above table, 250 lies between the cubes of 6 and 7. The *lower* of the two figures—in this case 6—will be the first figure of the cube root. The correct answer, therefore, is 63.

Once more example will make this clear. If the number called out is 19,683, the last digit, 3, indicates that the last digit of the cube root is 7. Discarding the final three digits leaves 19, which falls between the cubes of 2 and 3. Two is the lower number, therefore we arrive at a final cube root of 27.

Actually, a professional lightning calculator would probably memorize all the cubes of integers from 1 to 100, then use this information for calculating higher cubes. But the method just described makes it an easy and effective trick for the amateur. Oddly enough, there are even simpler rules for finding integral roots of powers higher than three. It is especially easy to find fifth roots because any number and its fifth power have the same final digit.

Adding a Fibonacci Series

A less-known stunt of rapid calculation is that of adding almost instantly any ten numbers of a Fibonacci series (that is, a series in which each number is the sum of the two preceding numbers). The trick may be presented as follows. The performer asks someone to jot down any two numbers he wishes. For illustration, let us suppose he chooses 8 and 5. He writes either number below the other, then is told to add them to obtain a third number. The third number is now added to the one above it to obtain a fourth, and this continues until there are ten numbers in a vertical column.

8

5

13

18

31

49

80

129

209

338

While these numbers are being recorded the performer keeps his back turned. After the ten numbers are written, he turns around, draws a line below the column and quickly writes the sum of all ten numbers. To obtain the sum, simply note the fourth number from the bottom and multiply it by 11, an operation easily performed in the head. In this case the number is 80, therefore the answer is 80 times 11, or 880. The trick was contributed by Royal V. Heath to *The Jinx*, No. 91

(1940). (See *American Mathematical Monthly*, Nov., 1947, for an article by A. L. Epstein in which he discusses the stunt as part of a more general problem.)

Prediction tricks and mind-reading tricks with numbers are usually interchangeable—that is, a trick that can be presented as a prediction may also be presented as mind-reading, and vice-versa. For example, suppose the performer knows in advance the outcome of a calculation which the spectator thinks he cannot know. The magician can dramatize this knowledge by writing the outcome in advance on a slip of paper, in which case he performs a prediction trick. Or he may pretend to read the spectator's mind after the result has been obtained, in which case he performs a mind-reading trick. (As a third alternative, he may pretend to obtain the answer by lightning calculation.) Most of the tricks now to be discussed will lend themselves to such alternate methods of presentation, but we shall not waste words in continually calling this to the reader's attention.

Predicting a Number

Perhaps the oldest of all prediction tricks is that of asking someone to think of a number, to perform certain operations upon it, then announce the final result. The result coincides with a previously written prediction. To take a trivial example, the spectator is asked to double his number, add 8, halve the result, then subtract the original number. The answer will be half of whatever number you tell him to add. In this example the added number is 8 so the final result will be 4. If the spectator had been asked to add 10, the final answer would have been 5.

A more interesting trick of this sort begins by asking a person to write down the year of his birth and add to it the year of some important event in his life. To this sum he then adds his age, and finally, the number of years since the important event took place. Few people realize that the total of these four numbers must always be twice the current year. This enables you, of course, to predict the total in advance.

Curry's Version

Magician Paul Curry, in his book SOMETHING BORROWED, SOMETHING NEW, 1940, suggested presenting this trick as follows. When the spectator writes down the year of his birth, you pretend to receive the number telepathically and to write it on your own sheet of paper without letting him see what you have written. Pretend to obtain his other three numbers in the same manner. Actually, you may write any numbers you please. While he is adding his four numbers and writes the sum beneath, you pretend to do the same, writing as your sum the number which you know will be his total. Now tell him that you do not wish anyone to see his age (if the spectator is of the fair sex, this concern will be even more appropriate), so suggest that he black out all four numbers with his pencil, leaving only the total. You do the same. The two sums are now compared and found to be identical. Such a presentation gives the impression that you somehow knew all four of his numbers, though of course you did not know any of them. It is an effective way to perform any number trick in which the answer is known in advance.

When you ask the person to put down his age, be sure to request that it be his age as of December 31 of that year. Otherwise he may have a birthday coming up before the year ends, in which case his total will be off by 1. Royal Heath, in MATHEMAGIC, suggests having the spectator also include in his sum an irrelevant figure such as the number of people in the room. Since this is known to you also, you merely add it to twice the current year to obtain the final answer. This serves to conceal the working of the trick. Also, should you have occasion to repeat the stunt, you can use a different figure (such as the day of the month) and end with a different answer.

Al Baker's Version

An interesting handling of the same trick was suggested by the New York magician Al Baker. You first ask a spectator to write the year of his birth without letting you see what he is writing. By observing the motions of his pencil it is not diffi-

cult to guess the last two figures of this date. Actually, you
need only be sure of the final digit because it is easy to guess
anyone's age within a ten-year period. At this point you may
turn your back while you direct him to add to the date of his
birth the date of an important event in his life. To this total
he adds the number of years that have elapsed since the important event. Because the last two figures always add to the
current year, you have only to add the current year to the year
of birth to obtain the final total. Performed in this way, the
total will of course be different each time you repeat the trick
with another person. Al Baker explained this trick in 1923
in a rare publication called *Al Baker's Complete Manuscript,*
which sold for fifty dollars during the twenties. This is the
earliest date I have found for a published explanation of the
principle on which the trick is based.

A similar type of trick is that of asking a spectator to
perform certain operations upon a thought-of number; he
announces his result, and from it you immediately tell him the
original number. Tricks of both types are found in the earliest
treatises on recreational mathematics. They are easy to invent
and scores of them have been recorded. The interested reader
will find representative samples in Ball's MATHEMATICAL REC-
REATIONS, Kraitchik's MATHEMATICAL RECREATIONS, and Heath's
MATHEMAGIC. The latter book is a collection of entertaining
number tricks by Royal V. Heath, first issued in 1933 and re-
issued by Dover Publications in 1953. RAINY DAY DIVERSIONS,
1907, by Carolyn Wells, also contains some excellent presentation ideas for arithmetical tricks of this nature.

Divining a Number

The most remarkable trick of this type has not to my knowledge been previously published. It differs from others of its
kind in the fact that at no time, during or after the series of
operations performed on a thought-of number, does the spectator give his results to the performer. However, from certain
clues obtained along the way, the magician is able to learn the
number.

The trick breaks down into the following steps:

1. Ask someone to think of a number from 1 to 10 inclusive.

2. Tell him to multiply it by 3.

3. Ask him to divide the result by 2.

4. At this point it is necessary for you to know whether he has a remainder of $\frac{1}{2}$. To gain this information, ask him to multiply by 3 once more. If he does so quickly, without hesitation, you can be reasonably certain that he does not have a fraction to deal with. If he does have a fraction, he will hesitate and look puzzled. He may even ask, "What about fractions?" In either case, if you suspect he has a fraction you say, "By the way, your last result has a fractional remainder, does it not? I thought so. Please eliminate the fraction entirely by making it the next whole number. For example, if your result is $10\frac{1}{2}$, eliminate the fraction by making it 11."

If there was a fractional remainder, you must remember the key number 1. If there was no remainder, you have nothing to recall.

5. After he has multiplied by 3, according to instructions given above, ask him to divide by 2 once more.

6. Again, you must learn whether he has a fractional remainder. To do this, say, "You now have, do you not, a whole number? That is, there is no fraction?" If he nods, say, "I thought so," and continue. If, however, he tells you that you are wrong, look puzzled a moment, then say, "Well, in that case get rid of the fraction by taking the next whole number."

If a fraction is involved at this point, remember the key number 2. Otherwise you remember nothing.

7. Ask him to add 2 to his result.

8. Tell him to subtract 11. These last two steps are the same, of course, as subtracting 9, but doing it this way serves to conceal the nine-principle involved.

9. If he tells you he is unable to subtract 11 because his last result is too small, you can immediately tell him the chosen number with which he started. If you are remembering only the key number 1, then the chosen number is 1. If you are remembering only the key number 2, the original number is 2. If you are recalling both key numbers, add them to obtain 3 as the answer.

Should he proceed with the subtraction of 11, you know that the thought-of number is higher than 3. Remember the key number 4 and continue as follows.

10. Ask him to add 2.

11. Tell him to subtract 11.

12. If he is unable to subtract 11, then total the key numbers you are remembering to obtain the answer.

If he says nothing and proceeds with the subtraction, then total the key numbers and add an additional 4 to obtain the answer.

The trick may seem unduly complicated, but if you go over it carefully you will soon become familiar with the procedure. Of course you can vary the subtractions of 9's in any way you wish. For example, instead of having him add 2 and take away 11, as explained above, you can tell him to add 5 and subtract 14, or add 1 and subtract 10.

After performing the trick several times you will discover how to give instructions in such a way that the spectator will not be aware of the fact that he is giving you clues about his original number. After a series of apparently meaningless arithmetical operations, and without telling you any of his results, he will be surprised to hear you name the number with which he started.

This trick was explained to me by the New York City amateur magician, Edmund Balducci, who in turn had been told the trick by a man now deceased, so the originator is not known. The trick is a combination of elements in two older tricks which may be found in the section titled "The Magic of Numbers" in THE MAGICIAN'S OWN BOOK, published in the mid-nineteenth century.

The Mysteries of Nine

The number 9 is a key figure in the trick just described. A great many other number tricks seem to exploit certain curious properties of 9. For example, if you reverse a three-digit number (provided the first and last digits are not alike) and subtract the smaller from the larger, the answer will always have a 9 as the center figure, and the two end digits will add to 9.

This means that if you are told either the first or last digit in the answer, you can immediately name the entire number.

If the answer is now reversed and the two numbers added, the sum will naturally be 1089. A popular number trick is to write 1089 in advance on a sheet of paper, placing it face down on the table. After the spectator has concluded the series of operations described above, and announced the final result as 1089, display your prediction by holding the sheet upside-down. It will read 6801 which is not, of course, the correct answer. Look perplexed for a moment, then apologize for holding the paper upside-down. Turn it around to reveal the correct total. This bit of by-play adds an entertaining touch to the presentation.

T. O'Connor Sloane, in his RAPID ARITHMETIC, 1922, suggested performing the trick with dollar and cent figures. Ask someone to put down a sum of money less than ten dollars and more than a dollar. The first and last figures must not be alike. The trick is then performed as described, resulting in a total of $10.89.

Digital Roots

If all the digits in a given number are added, then the digits in the sum added, and this continued until only a single digit remains, that digit is known as the *digital root* of the original number. The fastest way to arrive at a digital root is by a process called "casting out nines." Suppose, for example, we wish to know the digital root of 87,345,691. We first add the digits 8 and 7 to obtain 15, then immediately add the 5 and 1 to obtain 6. This is the same as subtracting or "casting out" 9 from 15. Six is now added to the next digit, 3, giving us 9. Nine plus 4 is 13, which we immediately reduce to the root of 4, and so on until we reach the end of the number. The digit 7, obtained in this way, will be the digital root of the entire series.

A great many number tricks are based on operations that seem to result in a random number, but actually end with a number that has a digital root of 9. When this occurs, the performer can ask the spectator to circle any digit (except zero) in the answer and call out the remaining digits in any order he pleases. The performer is then able to name the circled number. To do this he simply adds the numbers called out, casting out nines as he goes along, so that by the time the last digit is called he knows the digital root of the entire series. If this is 9, then he knows a 9 was circled. Otherwise, he subtracts it from 9 to obtain the circled digit.

Here are a few of the many procedures that result in a number with a digital root of 9.

(1) Write down any number, as large as desired, then "scramble" (rearrange) the digits in any desired manner to produce a second number. Subtract the smaller from the larger.

(2) Write down any number. Add all the digits and subtract the sum from the original number.

(3) Write down any number. Find the sum of its digits, multiply by 8 and add the result to the original number.

(4) Write down any number and multiply it by 9 or a multiple of 9. (All multiples of 9 have digital roots of 9 and conversely, all numbers with a digital root of 9 are multiples of 9.)

(5) Write down any number, add two scrambled forms of the same number to it, then square the result.

If you wish, you can further conceal the method by introducing random numbers and operations before you have the spectator perform the essential step. For instance, you may ask him to write down the amount of change in his pocket, multiply it by the number of people in the room, add the year of an important event in his life, etc., then multiply the final result by 9. The last step of course is the only one that matters. Once the number with a digital root of 9 is obtained, you can then have the spectator circle a digit and proceed with the effect as previously described.

Persistent Root

Take any number with a digital root of 9, scramble it to form a second number, scramble it again to form a third, and continue in this manner until you have as many numbers as you please. Add all these numbers together and the sum also will have a digital root of 9. Similarly, a number with a digital root of 9 can be multiplied by any number and the result will have a 9 root.

Many tricks can be built around these persistencies. For example, suppose you have obtained a dollar bill with a serial number that has a digital root of 9. Carry the bill with you until you wish to show the trick. Ask someone to jot down a series of random digits, then as an afterthought, take the bill from your pocket and suggest instead that he copy the serial number of the bill—a convenient way, you explain, of obtaining random numbers. He then scrambles the digits to obtain several new numbers, adds all the numbers without letting you see the calculation, multiplies the answer by any number he wishes, and finally circles one digit in the answer. If he now calls out the remaining digits, you can of course tell him the circled number.

Another novel presentation is to start with figures of the date on which you perform the trick—that is, the number of the month, the day of the month, and the year. For the year, you have a choice of using the last two figures of the year or all four figures. On approximately two out of every nine days you will find that by proper choice you can obtain a series of numbers with a digital root of 9. It is on these days that you perform the trick. For example, suppose the date is March 29, 1958. Ask someone to write the date as 3–29–58. Because this series has a digital root of 9, you can proceed exactly as in the dollar bill trick just described, or follow any other procedure that does not change the digital root.

Guessing Someone's Age

An interesting method of learning a person's age begins by asking him to perform any of the operations that produce a number with a digital root of 9. Tell him to add his age to this number and give you the total. From this total you can easily determine his age. First obtain the digital root of the total. Now keep adding 9 to it until you reach a number you estimate to be the closest to his age, and it will be his age. For example, suppose you ask a person to write down any number and multiply it by 9. He does this and arrives at the number 2,826. To this he adds his age of 40 and tells you that the result is 2,866. This has a root of 4. By adding 9's to 4 you obtain the series 13, 22, 31, 40, 49, etc. Since it is not difficult to estimate a person's age within nine years, you settle on 40 as the correct answer.

Accountants often check addition and multiplication problems by the use of digital roots. For example, addition may be checked by obtaining the digital root of all the digits in the numbers to be added, then comparing it with the digital root of the answer. If the answer is correct, these roots must be the same. This fact can be applied to a trick as follows.

An Addition Trick

Let someone form a problem of addition by writing a series of large numbers, one below another. With practice you should be able to cast out nines almost as rapidly as he writes the numbers, so that when he has completed the problem you know the digital root of the entire series. You now turn your back while he adds the numbers. If he then circles a digit (not a zero) in the answer and calls off the remaining digits, you can name the circled number. This is done by obtaining the digital root of the numbers called out, and subtracting it from the previous root which you are remembering. If the second root is larger than the first, add 9 to the first root before subtracting. If the two roots are the same, then of course a 9 was circled.

A Multiplication Trick

A similar trick can be performed with a problem of multiplication, owing to the fact that if the digital roots of two numbers are multiplied and the product reduced to a root, it will correspond to the digital root of the product of the two numbers. Thus you may ask someone to write a fairly large number, say of five or six digits, then underneath it write a second large number. As he does this, you obtain the digital root of each number, multiply the two roots, and reduce the answer to a root.

Now turn your back while he multiplies the two large numbers. Ask him to circle any number other than a zero in the answer, then call off to you the remaining digits in any order. As before, you obtain the circled number by subtracting the digital root of the numbers called from the digital root you are remembering. Again, if the second root is larger than the first, add 9 to the first root before you subtract.

The Mysteries of Seven

Both these tricks are discussed by the elder Sam Loyd in an interesting short article titled "Freaks in Figures," *Woman's Home Companion*, Nov., 1904. Loyd correctly points out that all the so-called "mysterious properties" of 9 arise from the simple fact that it is the last digit in our decimal number system. In a system of notation based on 8 instead of 10, the number 7 would acquire the same curious properties. We can easily verify this assertion. First let us list the numbers from 1 through 20 in a number system based on 8, together with their equivalents in a decimal system.

8 system	10 system
1	1
2	2
3	3
4	4
5	5
6	6
7	7
10	8
11	9
12	10
13	11
14	12
15	13
16	14
17	15
20	16

Now suppose we take the number 341 and subtract from it its reverse form of 143. We begin by taking 3 from 11. In our decimal system this is the same as taking 3 from 9, giving us an answer of 6. Six is the same symbol in both systems, hence it is the last digit of the answer. Continuing in this manner we obtain 176 as the complete answer.

$$341$$
$$143$$
$$\overline{}$$
$$176$$

You will observe that the center number is 7 and the two end numbers add to 7. This is exactly the same as what happens in the decimal equivalent of this trick, described earlier, except that 7 is the key number rather than 9.

Similar tests can be made of all the other tricks based on the properties of 9 in a decimal system and they will be found to have analogs in the other system, with 7 as the "mysterious" figure. By choosing an appropriate number system, we can transfer the magic properties to any number we desire. In this way we see that the properties do not spring from the intrinsic character of 9, but only from the fact that it is the last digit in our decimal system of numeration.

Confusing the properties of a number with properties acquired from its position in a number system is a common mistake. Thus at one time it was thought that 7, for some obscure reason, showed up with less than average frequency in the numbers which form the endless decimal of *pi*. "There is but one number which is treated with an unfairness which is incredible as an accident;" wrote Augustus De Morgan in his BUDGET OF PARADOXES, "and that number is the mysterious number *seven*!" De Morgan was not serious, of course; knowing full well that the digits in *pi* would be entirely different in another system of numeration. Actually, even in a decimal system the apparent paucity of sevens in *pi* proved to be due to errors in an early calculation by William Shanks. In 1873, after fifteen years of arduous labor, Shanks managed to calculate *pi* to 707 inaccurate decimals (his mistake on the 528th decimal threw off all subsequent digits). In 1949 the giant brain, ENIAC, took a weekend off from more important tasks and calculated *pi* to over 2,000 accurate decimals. No mysterious frequency deviations were found for any of the digits. (See N. T. Gridgeman's amusing article, "Circumetrics," in *The Scientific Monthly*, July, 1953.)

Predicting a Sum

Is it possible to know in advance the sum of a problem in addition, all the numbers of which are given at random by members of the audience?   Magicians have worked out many ingenious solutions to this problem that cannot be gone into here because they involve the use of stooges, sleight-of-hand, gimmicked slates, or other forms of nonmathematical deception.

If, however, the performer alternates with a spectator in supplying numbers for a sum, it is possible for him to bring the total to a desired number without recourse to other than purely mathematical means.   A simple and ancient method is as follows.   Suppose you desire to end with the sum 23,843. Remove the first digit, 2, and add it to the remaining number to obtain 3,845.   This is the first number that you write.

A spectator is now asked to write a four-digit number beneath it.

$$3845$$
$$1528$$

Below this you now write another four-digit number, apparently at random.   Actually, under each digit of the spectator's number you write its difference from 9.

$$3845$$
$$1528$$
$$8471$$

The spectator is asked to write another four-digit number. After he does so, you write a fifth number.   As before, you choose digits that add to 9.

Your number	3845	
His number	1528 ⎱	digits add to nine
Your number	8471 ⎰	
His number	2911 ⎱	digits add to nine
Your number	7088 ⎰	

When these five numbers are added, the total will be exactly 23,843.

In the example just given, the first digit of the predicted answer happened to be a 2. This means there must be *two pairs* of numbers, the digits of which add to 9, making five numbers in all to be added. If the first digit of the desired answer is a 3, then there must be *three* pairs of numbers adding to 9, and so on for the higher digits. In all cases, the first number you write is obtained by removing the first digit of the desired answer and adding it to what remains. The principle applies to numbers of any size provided all the separate additives have the same number of digits.

There are many variations of the trick. For example, you may ask a spectator to write the first number. You then write a number beneath it, making the digits add to 9. The spectator writes a third number. You write a fourth, using the nine-principle again. The spectator writes the fifth and last number, then you draw a line beneath the sum and immediately write the total. Or if you prefer, you may turn your back while the spectator totals the numbers, then tell him the answer without seeing what he has written. The answer is obtained, of course, by *subtracting* 2 from the fifth number and placing the 2 in front of the difference.

If you wish, you can make this a much longer problem of addition. For example—you and the spectator write six pairs of numbers, each pair adding to 99. . . . The spectator writes a final number, making 13 numbers in all. The answer is obtained by subtracting 6 from the thirteenth number and placing the 6 in front of the remainder. If the sum is continued until there are, say, 28 pairs of numbers before the final number, the principle is unchanged. Subtract 28 from the last number and place 28 in front of what remains.

Still another variation of the trick is to let the spectator himself write the prediction. Suppose he writes 538. Remove the 5 and add it to the remaining number to obtain 43. This is the first number you write. You now alternate pairs of two-digit numbers, using the nine-principle, until five of such pairs have been written below the first number.

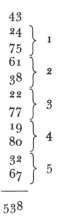

$$
\begin{array}{r}
43 \\
24 \\
75 \\
61 \\
38 \\
22 \\
77 \\
19 \\
80 \\
32 \\
67 \\
\hline
538
\end{array}
$$

The answer is of course the number predicted by the spectator.

Al Baker's "Numero"

The American magician Al Baker once worked out an amusing presentation of this trick, which he called "Numero" and published in *The Jinx*, July, 1936. In Baker's version, after revealing the predicted answer to a sum, the total is shown to be transposable into the spectator's first name.

The following alphabet code is involved.

A	1	I	9
B	2	J	0
C	3	K	1
D	4	L	2
E	5	M	3
F	6	N	4
G	7	O	5
H	8	P	6

Q — 7	V — 2
R — 8	W — 3
S — 9	X — 4
T — 0	Y — 5
U — 1	Z — 6

Let's assume the spectator's first name is Harry. Before showing the trick, consult the code to obtain the numerical equivalent of the letters in "Harry"—namely 81885. Place a 2 in front of this to make 281885. This is the prediction you write on a piece of paper and place aside for later reference.

The sum consists of five numbers. The first number, written by you, is 81887. This is arrived at by adding 2 to 81885. Harry writes a five-figure number beneath. You follow with a third number, using the nine-principle. He writes the fourth, and you add the last, again using the nine-principle. When he adds the five numbers the total will of course be the number you predicted.

The trick is apparently over, but you now proceed with an amusing climax. Cross out the first digit in the answer, leaving 81885. Write down the figure alphabet, circle the letters of "Harry," then show how these letters correspond to 81885. The trick can be used of course to produce any desired word or phrase. Crossing out the 2 at the beginning of the answer is a weak feature, but it is necessary if you want to limit the sum to five numbers.

Psychological Forces

A quite different category of prediction (or mind-reading) number tricks is based on what magicians call "psychological forces." These effects do not work every time, but for obscure psychological reasons the odds in favor of their success are higher than one would suspect. A simple example is the tendency of most people to select the number 7 if asked to name a number from 1 to 10, or the number 3 if you ask for a number between 1 and 5.

A remarkable psychological force, the inventor of which is not known to me, works as follows. Write the number 37 on a slip of paper and place it face down. Now say to someone,

"I want you to name a two-digit number between 1 and 50. Both digits must be *odd*, and they must not be alike. For instance, you cannot name 11." Curiously, the odds are good that he will name 37. (The second most likely choice seems to be 35.) Actually, his choice is restricted to only eight numbers. Your mention of 11 tends to send his mind into the thirties where 37 is apparently the number most often selected.

If you succeed in this trick, try following it with another one by asking for a two-digit number between 50 and 100, both digits *even*. Again, the digits must not be the same. Here the spectator's choice is restricted to six numbers, of which 68 seems to be chosen most often. If playing cards are handy, you may make your prediction by placing a six and an eight face down on the table. This increases your chances of success because you can display two possible answers, 68 and 86, depending on which card you turn up first.

Index of Names

175

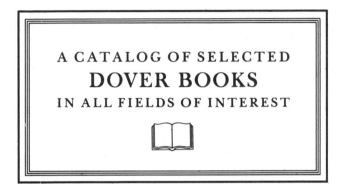

A CATALOG OF SELECTED
DOVER BOOKS
IN ALL FIELDS OF INTEREST

A CATALOG OF SELECTED
DOVER BOOKS
IN ALL FIELDS OF INTEREST

DRAWINGS OF REMBRANDT, edited by Seymour Slive. Updated Lippmann, Hofstede de Groot edition, with definitive scholarly apparatus. All portraits, biblical sketches, landscapes, nudes. Oriental figures, classical studies, together with selection of work by followers. 550 illustrations. Total of 630pp. 9⅜ × 12¼.
21485-0, 21486-9 Pa., Two-vol. set $29.90

GHOST AND HORROR STORIES OF AMBROSE BIERCE, Ambrose Bierce. 24 tales vividly imagined, strangely prophetic, and decades ahead of their time in technical skill: "The Damned Thing," "An Inhabitant of Carcosa," "The Eyes of the Panther," "Moxon's Master," and 20 more. 199pp. 5⅜ × 8½. 20767-6 Pa. $4.95

ETHICAL WRITINGS OF MAIMONIDES, Maimonides. Most significant ethical works of great medieval sage, newly translated for utmost precision, readability. Laws Concerning Character Traits, Eight Chapters, more. 192pp. 5⅜ × 8½.
24522-5 Pa. $5.95

THE EXPLORATION OF THE COLORADO RIVER AND ITS CANYONS, J. W. Powell. Full text of Powell's 1,000-mile expedition down the fabled Colorado in 1869. Superb account of terrain, geology, vegetation, Indians, famine, mutiny, treacherous rapids, mighty canyons, during exploration of last unknown part of continental U.S. 400pp. 5⅜ × 8½. 20094-9 Pa. $8.95

HISTORY OF PHILOSOPHY, Julián Marías. Clearest one-volume history on the market. Every major philosopher and dozens of others, to Existentialism and later. 505pp. 5⅜ × 8½. 21739-6 Pa. $9.95

ALL ABOUT LIGHTNING, Martin A. Uman. Highly readable nontechnical survey of nature and causes of lightning, thunderstorms, ball lightning, St. Elmo's Fire, much more. Illustrated. 192pp. 5⅜ × 8½. 25237-X Pa. $5.95

SAILING ALONE AROUND THE WORLD, Captain Joshua Slocum. First man to sail around the world, alone, in small boat. One of great feats of seamanship told in delightful manner. 67 illustrations. 294pp. 5⅜ × 8½. 20326-3 Pa. $4.95

LETTERS AND NOTES ON THE MANNERS, CUSTOMS AND CONDITIONS OF THE NORTH AMERICAN INDIANS, George Catlin. Classic account of life among Plains Indians: ceremonies, hunt, warfare, etc. 312 plates. 572pp. of text. 6⅛ × 9¼. 22118-0, 22119-9, Pa., Two-vol. set $17.90

THE SECRET LIFE OF SALVADOR DALÍ, Salvador Dalí. Outrageous but fascinating autobiography through Dalí's thirties with scores of drawings and sketches and 80 photographs. A must for lovers of 20th-century art. 432pp. 6½ × 9¼. (Available in U.S. only) 27454-3 Pa. $9.95

CATALOG OF DOVER BOOKS

THE BOOK OF BEASTS: Being a Translation from a Latin Bestiary of the Twelfth Century, T. H. White. Wonderful catalog of real and fanciful beasts: manticore, griffin, phoenix, amphivius, jaculus, many more. White's witty erudite commentary on scientific, historical aspects enhances fascinating glimpse of medieval mind. Illustrated. 296pp. 5⅜ × 8¼. (Available in U.S. only)   24609-4 Pa. $7.95

FRANK LLOYD WRIGHT: Architecture and Nature with 160 Illustrations, Donald Hoffmann. Profusely illustrated study of influence of nature—especially prairie—on Wright's designs for Fallingwater, Robie House, Guggenheim Museum, other masterpieces. 96pp. 9¼ × 10¾.   25098-9 Pa. $8.95

LIMBERT ARTS AND CRAFTS FURNITURE: The Complete 1903 Catalog, Charles P. Limbert and Company. Rare catalog depicting 188 pieces of Mission-style furniture: fold-down tables and desks, bookcases, library and octagonal tables, chairs, more. Descriptive captions. 80pp. 9⅜ × 12¼.   27120-X Pa. $6.95

YEARS WITH FRANK LLOYD WRIGHT: Apprentice to Genius, Edgar Tafel. Insightful memoir by a former apprentice presents a revealing portrait of Wright the man, the inspired teacher, the greatest American architect. 372 black-and-white illustrations. Preface. Index. vi + 228pp. 8¼ × 11.   24801-1 Pa. $10.95

THE STORY OF KING ARTHUR AND HIS KNIGHTS, Howard Pyle. Enchanting version of King Arthur fable has delighted generations with imaginative narratives of exciting adventures and unforgettable illustrations by the author. 41 illustrations. xviii + 313pp. 6⅛ × 9¼.   21445-1 Pa. $6.95

THE GODS OF THE EGYPTIANS, E. A. Wallis Budge. Thorough coverage of numerous gods of ancient Egypt by foremost Egyptologist. Information on evolution of cults, rites and gods; the cult of Osiris; the Book of the Dead and its rites; the sacred animals and birds; Heaven and Hell; and more. 956pp. 6⅛ × 9¼.
22055-9, 22056-7 Pa., Two-vol. set $22.90

A THEOLOGICO-POLITICAL TREATISE, Benedict Spinoza. Also contains unfinished *Political Treatise*. Great classic on religious liberty, theory of government on common consent. R. Elwes translation. Total of 421pp. 5⅜ × 8½.
20249-6 Pa. $7.95

INCIDENTS OF TRAVEL IN CENTRAL AMERICA, CHIAPAS, AND YUCATAN, John L. Stephens. Almost single-handed discovery of Maya culture; exploration of ruined cities, monuments, temples; customs of Indians. 115 drawings. 892pp. 5⅜ × 8½.   22404-X, 22405-8 Pa., Two-vol. set $17.90

LOS CAPRICHOS, Francisco Goya. 80 plates of wild, grotesque monsters and caricatures. Prado manuscript included. 183pp. 6⅜ × 9⅜.   22384-1 Pa. $6.95

AUTOBIOGRAPHY: The Story of My Experiments with Truth, Mohandas K. Gandhi. Not hagiography, but Gandhi in his own words. Boyhood, legal studies, purification, the growth of the Satyagraha (nonviolent protest) movement. Critical, inspiring work of the man who freed India. 480pp. 5⅜ × 8½. (Available in U.S. only)
24593-4 Pa. $6.95

ILLUSTRATED DICTIONARY OF HISTORIC ARCHITECTURE, edited by Cyril M. Harris. Extraordinary compendium of clear, concise definitions for over 5,000 important architectural terms complemented by over 2,000 line drawings. Covers full spectrum of architecture from ancient ruins to 20th-century Modernism. Preface. 592pp. 7½ × 9⅜. 24444-X Pa. $15.95

THE NIGHT BEFORE CHRISTMAS, Clement C. Moore. Full text, and woodcuts from original 1848 book. Also critical, historical material. 19 illustrations. 40pp. 4⅝ × 6. 22797-9 Pa. $2.50

THE LESSON OF JAPANESE ARCHITECTURE: 165 Photographs, Jiro Harada. Memorable gallery of 165 photographs taken in the 1930s of exquisite Japanese homes of the well-to-do and historic buildings. 13 line diagrams. 192pp. 8⅜ × 11¼. 24778-3 Pa. $10.95

THE AUTOBIOGRAPHY OF CHARLES DARWIN AND SELECTED LETTERS, edited by Francis Darwin. The fascinating life of eccentric genius composed of an intimate memoir by Darwin (intended for his children); commentary by his son, Francis; hundreds of fragments from notebooks, journals, papers; and letters to and from Lyell, Hooker, Huxley, Wallace and Henslow. xi + 365pp. 5⅜ × 8. 20479-0 Pa. $6.95

WONDERS OF THE SKY: Observing Rainbows, Comets, Eclipses, the Stars and Other Phenomena, Fred Schaaf. Charming, easy-to-read poetic guide to all manner of celestial events visible to the naked eye. Mock suns, glories, Belt of Venus, more. Illustrated. 299pp. 5¼ × 8¼. 24402-4 Pa. $8.95

BURNHAM'S CELESTIAL HANDBOOK, Robert Burnham, Jr. Thorough guide to the stars beyond our solar system. Exhaustive treatment. Alphabetical by constellation: Andromeda to Cetus in Vol. 1; Chamaeleon to Orion in Vol. 2; and Pavo to Vulpecula in Vol. 3. Hundreds of illustrations. Index in Vol. 3. 2,000pp. 6⅛ × 9¼. 23567-X, 23568-8, 23673-0 Pa., Three-vol. set $41.85

STAR NAMES: Their Lore and Meaning, Richard Hinckley Allen. Fascinating history of names various cultures have given to constellations and literary and folkloristic uses that have been made of stars. Indexes to subjects. Arabic and Greek names. Biblical references. Bibliography. 563pp. 5⅜ × 8½. 21079-0 Pa. $9.95

THIRTY YEARS THAT SHOOK PHYSICS: The Story of Quantum Theory, George Gamow. Lucid, accessible introduction to influential theory of energy and matter. Careful explanations of Dirac's anti-particles, Bohr's model of the atom, much more. 12 plates. Numerous drawings. 240pp. 5⅜ × 8½. 24895-X Pa. $6.95

CHINESE DOMESTIC FURNITURE IN PHOTOGRAPHS AND MEASURED DRAWINGS, Gustav Ecke. A rare volume, now affordably priced for antique collectors, furniture buffs and art historians. Detailed review of styles ranging from early Shang to late Ming. Unabridged republication. 161 black-and-white drawings, photos. Total of 224pp. 8⅜ × 11¼. (Available in U.S. only) 25171-3 Pa. $14.95

VINCENT VAN GOGH: A Biography, Julius Meier-Graefe. Dynamic, penetrating study of artist's life, relationship with brother, Theo, painting techniques, travels, more. Readable, engrossing. 160pp. 5⅜ × 8½. (Available in U.S. only) 25253-1 Pa. $4.95

HOW TO WRITE, Gertrude Stein. Gertrude Stein claimed anyone could understand her unconventional writing—here are clues to help. Fascinating improvisations, language experiments, explanations illuminate Stein's craft and the art of writing. Total of 414pp. 4⅝ × 6⅜. 23144-5 Pa. $6.95

ADVENTURES AT SEA IN THE GREAT AGE OF SAIL: Five Firsthand Narratives, edited by Elliot Snow. Rare true accounts of exploration, whaling, shipwreck, fierce natives, trade, shipboard life, more. 33 illustrations. Introduction. 353pp. 5⅜ × 8½. 25177-2 Pa. $9.95

THE HERBAL OR GENERAL HISTORY OF PLANTS, John Gerard. Classic descriptions of about 2,850 plants—with over 2,700 illustrations—includes Latin and English names, physical descriptions, varieties, time and place of growth, more. 2,706 illustrations. xlv + 1,678pp. 8½ × 12¼. 23147-X Cloth. $89.95

DOROTHY AND THE WIZARD IN OZ, L. Frank Baum. Dorothy and the Wizard visit the center of the Earth, where people are vegetables, glass houses grow and Oz characters reappear. Classic sequel to *Wizard of Oz*. 256pp. 5⅜ × 8.
24714-7 Pa. $5.95

SONGS OF EXPERIENCE: Facsimile Reproduction with 26 Plates in Full Color, William Blake. This facsimile of Blake's original "Illuminated Book" reproduces 26 full-color plates from a rare 1826 edition. Includes "The Tyger," "London," "Holy Thursday," and other immortal poems. 26 color plates. Printed text of poems. 48pp. 5¼ × 7. 24636-1 Pa. $3.95

SONGS OF INNOCENCE, William Blake. The first and most popular of Blake's famous "Illuminated Books," in a facsimile edition reproducing all 31 brightly colored plates. Additional printed text of each poem. 64pp. 5¼ × 7.
22764-2 Pa. $3.95

PRECIOUS STONES, Max Bauer. Classic, thorough study of diamonds, rubies, emeralds, garnets, etc.: physical character, occurrence, properties, use, similar topics. 20 plates, 8 in color. 94 figures. 659pp. 6⅛ × 9¼.
21910-0, 21911-9 Pa., Two-vol. set $21.90

ENCYCLOPEDIA OF VICTORIAN NEEDLEWORK, S. F. A. Caulfeild and Blanche Saward. Full, precise descriptions of stitches, techniques for dozens of needlecrafts—most exhaustive reference of its kind. Over 800 figures. Total of 679pp. 8⅛ × 11. 22800-2, 22801-0 Pa., Two-vol. set $26.90

THE MARVELOUS LAND OF OZ, L. Frank Baum. Second Oz book, the Scarecrow and Tin Woodman are back with hero named Tip, Oz magic. 136 illustrations. 287pp. 5⅜ × 8½. 20692-0 Pa. $5.95

WILD FOWL DECOYS, Joel Barber. Basic book on the subject, by foremost authority and collector. Reveals history of decoy making and rigging, place in American culture, different kinds of decoys, how to make them, and how to use them. 140 plates. 156pp. 7⅞ × 10¾. 20011-6 Pa. $14.95

HISTORY OF LACE, Mrs. Bury Palliser. Definitive, profusely illustrated chronicle of lace from earliest times to late 19th century. Laces of Italy, Greece, England, France, Belgium, etc. Landmark of needlework scholarship. 266 illustrations. 672pp. 6⅛ × 9¼. 24742-2 Pa. $16.95

ILLUSTRATED GUIDE TO SHAKER FURNITURE, Robert Meader. All furniture and appurtenances, with much on unknown local styles. 235 photos. 146pp. 9 × 12. 22819-3 Pa. $9.95

WHALE SHIPS AND WHALING: A Pictorial Survey, George Francis Dow. Over 200 vintage engravings, drawings, photographs of barks, brigs, cutters, other vessels. Also harpoons, lances, whaling guns, many other artifacts. Comprehensive text by foremost authority. 207 black-and-white illustrations. 288pp. 6 × 9. 24808-9 Pa. $9.95

THE BERTRAMS, Anthony Trollope. Powerful portrayal of blind self-will and thwarted ambition includes one of Trollope's most heartrending love stories. 497pp. 5⅜ × 8½. 25119-5 Pa. $9.95

ADVENTURES WITH A HAND LENS, Richard Headstrom. Clearly written guide to observing and studying flowers and grasses, fish scales, moth and insect wings, egg cases, buds, feathers, seeds, leaf scars, moss, molds, ferns, common crystals, etc.—all with an ordinary, inexpensive magnifying glass. 209 exact line drawings aid in your discoveries. 220pp. 5⅜ × 8½. 23330-8 Pa. $5.95

RODIN ON ART AND ARTISTS, Auguste Rodin. Great sculptor's candid, wide-ranging comments on meaning of art; great artists; relation of sculpture to poetry, painting, music; philosophy of life, more. 76 superb black-and-white illustrations of Rodin's sculpture, drawings and prints. 119pp. 8⅜ × 11¼. 24487-3 Pa. $7.95

FIFTY CLASSIC FRENCH FILMS, 1912–1982: A Pictorial Record, Anthony Slide. Memorable stills from Grand Illusion, Beauty and the Beast, Hiroshima, Mon Amour, many more. Credits, plot synopses, reviews, etc. 160pp. 8¼ × 11. 25256-6 Pa. $11.95

THE PRINCIPLES OF PSYCHOLOGY, William James. Famous long course complete, unabridged. Stream of thought, time perception, memory, experimental methods; great work decades ahead of its time. 94 figures. 1,391pp. 5⅜ × 8½. 20381-6, 20382-4 Pa., Two-vol. set $25.90

BODIES IN A BOOKSHOP, R. T. Campbell. Challenging mystery of blackmail and murder with ingenious plot and superbly drawn characters. In the best tradition of British suspense fiction. 192pp. 5⅜ × 8½. 24720-1 Pa. $5.95

CALLAS: Portrait of a Prima Donna, George Jellinek. Renowned commentator on the musical scene chronicles incredible career and life of the most controversial, fascinating, influential operatic personality of our time. 64 black-and-white photographs. 416pp. 5⅜ × 8¼. 25047-4 Pa. $8.95

GEOMETRY, RELATIVITY AND THE FOURTH DIMENSION, Rudolph Rucker. Exposition of fourth dimension, concepts of relativity as Flatland characters continue adventures. Popular, easily followed yet accurate, profound. 141 illustrations. 133pp. 5⅜ × 8½. 23400-2 Pa. $4.95

HOUSEHOLD STORIES BY THE BROTHERS GRIMM, with pictures by Walter Crane. 53 classic stories—Rumpelstiltskin, Rapunzel, Hansel and Gretel, the Fisherman and his Wife, Snow White, Tom Thumb, Sleeping Beauty, Cinderella, and so much more—lavishly illustrated with original 19th-century drawings. 114 illustrations. x + 269pp. 5⅜ × 8½. 21080-4 Pa. $4.95

SUNDIALS, Albert Waugh. Far and away the best, most thorough coverage of ideas, mathematics concerned, types, construction, adjusting anywhere. Over 100 illustrations. 230pp. 5⅜ × 8½. 22947-5 Pa. $5.95

PICTURE HISTORY OF THE NORMANDIE: With 190 Illustrations, Frank O. Braynard. Full story of legendary French ocean liner: Art Deco interiors, design innovations, furnishings, celebrities, maiden voyage, tragic fire, much more. Extensive text. 144pp. 8⅞ × 11¾. 25257-4 Pa. $11.95

THE FIRST AMERICAN COOKBOOK: A Facsimile of "American Cookery," 1796, Amelia Simmons. Facsimile of the first American-written cookbook published in the United States contains authentic recipes for colonial favorites— pumpkin pudding, winter squash pudding, spruce beer, Indian slapjacks, and more. Introductory Essay and Glossary of colonial cooking terms. 80pp. 5⅜ × 8½.
24710-4 Pa. $3.50

101 PUZZLES IN THOUGHT AND LOGIC, C. R. Wylie, Jr. Solve murders and robberies, find out which fishermen are liars, how a blind man could possibly identify a color—purely by your own reasoning! 107pp. 5⅜ × 8½. 20367-0 Pa. $2.95

ANCIENT EGYPTIAN MYTHS AND LEGENDS, Lewis Spence. Examines animism, totemism, fetishism, creation myths, deities, alchemy, art and magic, other topics. Over 50 illustrations. 432pp. 5⅜ × 8½. 26525-0 Pa. $8.95

ANTHROPOLOGY AND MODERN LIFE, Franz Boas. Great anthropologist's classic treatise on race and culture. Introduction by Ruth Bunzel. Only inexpensive paperback edition. 255pp. 5⅜ × 8½. 25245-0 Pa. $7.95

THE TALE OF PETER RABBIT, Beatrix Potter. The inimitable Peter's terrifying adventure in Mr. McGregor's garden, with all 27 wonderful, full-color Potter illustrations. 55pp. 4¼ × 5½. 22827-4 Pa. $1.75

THREE PROPHETIC SCIENCE FICTION NOVELS, H. G. Wells. *When the Sleeper Wakes, A Story of the Days to Come* and *The Time Machine* (full version). 335pp. 5⅜ × 8½. (Available in U.S. only) 20605-X Pa. $8.95

APICIUS COOKERY AND DINING IN IMPERIAL ROME, edited and translated by Joseph Dommers Vehling. Oldest known cookbook in existence offers readers a clear picture of what foods Romans ate, how they prepared them, etc. 49 illustrations. 301pp. 6⅛ × 9¼. 23563-7 Pa. $8.95

SHAKESPEARE LEXICON AND QUOTATION DICTIONARY, Alexander Schmidt. Full definitions, locations, shades of meaning of every word in plays and poems. More than 50,000 exact quotations. 1,485pp. 6½ × 9¼.
22726-X, 22727-8 Pa., Two-vol. set $31.90

THE WORLD'S GREAT SPEECHES, edited by Lewis Copeland and Lawrence W. Lamm. Vast collection of 278 speeches from Greeks to 1970. Powerful and effective models; unique look at history. 842pp. 5⅜ × 8½. 20468-5 Pa. $12.95

THE BLUE FAIRY BOOK, Andrew Lang. The first, most famous collection, with many familiar tales: Little Red Riding Hood, Aladdin and the Wonderful Lamp, Puss in Boots, Sleeping Beauty, Hansel and Gretel, Rumpelstiltskin; 37 in all. 138 illustrations. 390pp. 5⅜ × 8½. 21437-0 Pa. $6.95

THE STORY OF THE CHAMPIONS OF THE ROUND TABLE, Howard Pyle. Sir Launcelot, Sir Tristram and Sir Percival in spirited adventures of love and triumph retold in Pyle's inimitable style. 50 drawings, 31 full-page. xviii + 329pp. 6½ × 9¼. 21883-X Pa. $7.95

THE MYTHS OF THE NORTH AMERICAN INDIANS, Lewis Spence. Myths and legends of the Algonquins, Iroquois, Pawnees and Sioux with comprehensive historical and ethnological commentary. 36 illustrations. 5⅜ × 8½.
25967-6 Pa. $8.95

GREAT DINOSAUR HUNTERS AND THEIR DISCOVERIES, Edwin H. Colbert. Fascinating, lavishly illustrated chronicle of dinosaur research, 1820s to 1960. Achievements of Cope, Marsh, Brown, Buckland, Mantell, Huxley, many others. 384pp. 5¼ × 8¼. 24701-5 Pa. $8.95

THE TASTEMAKERS, Russell Lynes. Informal, illustrated social history of American taste 1850s–1950s. First popularized categories Highbrow, Lowbrow, Middlebrow. 129 illustrations. New (1979) afterword. 384pp. 6 × 9.
23993-4 Pa. $8.95

NORTH AMERICAN INDIAN LIFE: Customs and Traditions of 23 Tribes, Elsie Clews Parsons (ed.). 27 fictionalized essays by noted anthropologists examine religion, customs, government, additional facets of life among the Winnebago, Crow, Zuni, Eskimo, other tribes. 480pp. 6⅛ × 9¼. 27377-6 Pa. $10.95

AUTHENTIC VICTORIAN DECORATION AND ORNAMENTATION IN FULL COLOR: 46 Plates from "Studies in Design," Christopher Dresser. Superb full-color lithographs reproduced from rare original portfolio of a major Victorian designer. 48pp. 9¼ × 12¼. 25083-0 Pa. $7.95

PRIMITIVE ART, Franz Boas. Remains the best text ever prepared on subject, thoroughly discussing Indian, African, Asian, Australian, and, especially, Northern American primitive art. Over 950 illustrations show ceramics, masks, totem poles, weapons, textiles, paintings, much more. 376pp. 5⅜ × 8. 20025-6 Pa. $8.95

SIDELIGHTS ON RELATIVITY, Albert Einstein. Unabridged republication of two lectures delivered by the great physicist in 1920–21. *Ether and Relativity* and *Geometry and Experience*. Elegant ideas in nonmathematical form, accessible to intelligent layman. vi + 56pp. 5⅜ × 8½. 24511-X Pa. $3.95

THE WIT AND HUMOR OF OSCAR WILDE, edited by Alvin Redman. More than 1,000 ripostes, paradoxes, wisecracks: Work is the curse of the drinking classes, I can resist everything except temptation, etc. 258pp. 5⅜ × 8½. 20602-5 Pa. $4.95

ADVENTURES WITH A MICROSCOPE, Richard Headstrom. 59 adventures with clothing fibers, protozoa, ferns and lichens, roots and leaves, much more. 142 illustrations. 232pp. 5⅜ × 8½. 23471-1 Pa. $4.95

PLANTS OF THE BIBLE, Harold N. Moldenke and Alma L. Moldenke. Standard reference to all 230 plants mentioned in Scriptures. Latin name, biblical reference, uses, modern identity, much more. Unsurpassed encyclopedic resource for scholars, botanists, nature lovers, students of Bible. Bibliography. Indexes. 123 black-and-white illustrations. 384pp. 6 × 9. 25069-5 Pa. $9.95

FAMOUS AMERICAN WOMEN: A Biographical Dictionary from Colonial Times to the Present, Robert McHenry, ed. From Pocahontas to Rosa Parks, 1,035 distinguished American women documented in separate biographical entries. Accurate, up-to-date data, numerous categories, spans 400 years. Indices. 493pp. 6½ × 9¼. 24523-3 Pa. $11.95

THE FABULOUS INTERIORS OF THE GREAT OCEAN LINERS IN HISTORIC PHOTOGRAPHS, William H. Miller, Jr. Some 200 superb photographs capture exquisite interiors of world's great "floating palaces"—1890s to 1980s: *Titanic, Ile de France, Queen Elizabeth, United States, Europa,* more. Approx. 200 black-and-white photographs. Captions. Text. Introduction. 160pp. 8⅞ × 11¾. 24756-2 Pa. $10.95

THE GREAT LUXURY LINERS, 1927–1954: A Photographic Record, William H. Miller, Jr. Nostalgic tribute to heyday of ocean liners. 186 photos of *Ile de France, Normandie, Leviathan, Queen Elizabeth, United States,* many others. Interior and exterior views. Introduction. Captions. 160pp. 9 × 12. 24056-8 Pa. $12.95

A NATURAL HISTORY OF THE DUCKS, John Charles Phillips. Great landmark of ornithology offers complete detailed coverage of nearly 200 species and subspecies of ducks: gadwall, sheldrake, merganser, pintail, many more. 74 full-color plates, 102 black-and-white. Bibliography. Total of 1,920pp. 8⅝ × 11¼. 25141-1, 25142-X Cloth., Two-vol. set $100.00

THE COMPLETE "MASTERS OF THE POSTER": All 256 Color Plates from "Les Maîtres de l'Affiche", Stanley Appelbaum (ed.). The most famous compilation ever made of the art of the great age of the poster, featuring works by Chéret, Steinlen, Toulouse-Lautrec, nearly 100 other artists. One poster per page. 272pp. 9¼ × 12¼. 26309-6 Pa. $29.95

THE TEN BOOKS OF ARCHITECTURE: The 1755 Leoni Edition, Leon Battista Alberti. Rare classic helped introduce the glories of ancient architecture to the Renaissance. 68 black-and-white plates. 336pp. 8⅜ × 11¼. 25239-6 Pa. $14.95

MISS MACKENZIE, Anthony Trollope. Minor masterpieces by Victorian master unmasks many truths about life in 19th-century England. First inexpensive edition in years. 392pp. 5⅜ × 8½. 25201-9 Pa. $8.95

THE RIME OF THE ANCIENT MARINER, Gustave Doré, Samuel Taylor Coleridge. Dramatic engravings considered by many to be his greatest work. The terrifying space of the open sea, the storms and whirlpools of an unknown ocean, the ice of Antarctica, more—all rendered in a powerful, chilling manner. Full text. 38 plates. 77pp. 9¼ × 12. 22305-1 Pa. $4.95

THE EXPEDITIONS OF ZEBULON MONTGOMERY PIKE, Zebulon Montgomery Pike. Fascinating firsthand accounts (1805–6) of exploration of Mississippi River, Indian wars, capture by Spanish dragoons, much more. 1,088pp. 5⅜ × 8½. 25254-X, 25255-8 Pa., Two-vol. set $25.90

A CONCISE HISTORY OF PHOTOGRAPHY: Third Revised Edition, Helmut Gernsheim. Best one-volume history—camera obscura, photochemistry, daguerreotypes, evolution of cameras, film, more. Also artistic aspects—landscape, portraits, fine art, etc. 281 black-and-white photographs. 26 in color. 176pp. 8⅜ × 11¼.
25128-4 Pa. $14.95

THE DORÉ BIBLE ILLUSTRATIONS, Gustave Doré. 241 detailed plates from the Bible: the Creation scenes, Adam and Eve, Flood, Babylon, battle sequences, life of Jesus, etc. Each plate is accompanied by the verses from the King James version of the Bible. 241pp. 9 × 12.
23004-X Pa. $9.95

WANDERINGS IN WEST AFRICA, Richard F. Burton. Great Victorian scholar/adventurer's invaluable descriptions of African tribal rituals, fetishism, culture, art, much more. Fascinating 19th-century account. 624pp. 5⅜ × 8½. 26890-X Pa. $12.95

HISTORIC HOMES OF THE AMERICAN PRESIDENTS, Second Revised Edition, Irvin Haas. Guide to homes occupied by every president from Washington to Bush. Visiting hours, travel routes, more. 175 photos. 160pp. 8¼ × 11.
26751-2 Pa. $9.95

THE HISTORY OF THE LEWIS AND CLARK EXPEDITION, Meriwether Lewis and William Clark, edited by Elliott Coues. Classic edition of Lewis and Clark's day-by-day journals that later became the basis for U.S. claims to Oregon and the West. Accurate and invaluable geographical, botanical, biological, meteorological and anthropological material. Total of 1,508pp. 5⅜ × 8½.
21268-8, 21269-6, 21270-X Pa., Three-vol. set $29.85

LANGUAGE, TRUTH AND LOGIC, Alfred J. Ayer. Famous, clear introduction to Vienna, Cambridge schools of Logical Positivism. Role of philosophy, elimination of metaphysics, nature of analysis, etc. 160pp. 5⅜ × 8½. (Available in U.S. and Canada only)
20010-8 Pa. $3.95

MATHEMATICS FOR THE NONMATHEMATICIAN, Morris Kline. Detailed, college-level treatment of mathematics in cultural and historical context, with numerous exercises. For liberal arts students. Preface. Recommended Reading Lists. Tables. Index. Numerous black-and-white figures. xvi + 641pp. 5⅜ × 8½.
24823-2 Pa. $11.95

HANDBOOK OF PICTORIAL SYMBOLS, Rudolph Modley. 3,250 signs and symbols, many systems in full; official or heavy commercial use. Arranged by subject. Most in Pictorial Archive series. 143pp. 8⅜ × 11. 23357-X Pa. $8.95

INCIDENTS OF TRAVEL IN YUCATAN, John L. Stephens. Classic (1843) exploration of jungles of Yucatan, looking for evidences of Maya civilization. Travel adventures, Mexican and Indian culture, etc. Total of 669pp. 5⅜ × 8½.
20926-1, 20927-X Pa., Two-vol. set $13.90

DEGAS: An Intimate Portrait, Ambroise Vollard. Charming, anecdotal memoir by famous art dealer of one of the greatest 19th-century French painters. 14 black-and-white illustrations. Introduction by Harold L. Van Doren. 96pp. 5⅜ × 8½.
25131-4 Pa. $4.95

PERSONAL NARRATIVE OF A PILGRIMAGE TO AL-MADINAH AND MECCAH, Richard F. Burton. Great travel classic by remarkably colorful personality. Burton, disguised as a Moroccan, visited sacred shrines of Islam, narrowly escaping death. 47 illustrations. 959pp. 5⅜ × 8½.
21217-3, 21218-1 Pa., Two-vol. set $19.90

PHRASE AND WORD ORIGINS, A. H. Holt. Entertaining, reliable, modern study of more than 1,200 colorful words, phrases, origins and histories. Much unexpected information. 254pp. 5⅜ × 8½.
20758-7 Pa. $5.95

THE RED THUMB MARK, R. Austin Freeman. In this first Dr. Thorndyke case, the great scientific detective draws fascinating conclusions from the nature of a single fingerprint. Exciting story, authentic science. 320pp. 5⅜ × 8½. (Available in U.S. only)
25210-8 Pa. $6.95

AN EGYPTIAN HIEROGLYPHIC DICTIONARY, E. A. Wallis Budge. Monumental work containing about 25,000 words or terms that occur in texts ranging from 3000 B.C. to 600 A.D. Each entry consists of a transliteration of the word, the word in hieroglyphs, and the meaning in English. 1,314pp. 6⅜ × 10.
23615-3, 23616-1 Pa., Two-vol. set $35.90

THE COMPLEAT STRATEGYST: Being a Primer on the Theory of Games of Strategy, J. D. Williams. Highly entertaining classic describes, with many illustrated examples, how to select best strategies in conflict situations. Prefaces. Appendices. xvi + 268pp. 5⅜ × 8½.
25101-2 Pa. $7.95

THE ROAD TO OZ, L. Frank Baum. Dorothy meets the Shaggy Man, little Button-Bright and the Rainbow's beautiful daughter in this delightful trip to the magical Land of Oz. 272pp. 5⅜ × 8.
25208-6 Pa. $5.95

POINT AND LINE TO PLANE, Wassily Kandinsky. Seminal exposition of role of point, line, other elements in nonobjective painting. Essential to understanding 20th-century art. 127 illustrations. 192pp. 6½ × 9¼.
23808-3 Pa. $5.95

LADY ANNA, Anthony Trollope. Moving chronicle of Countess Lovel's bitter struggle to win for herself and daughter Anna their rightful rank and fortune—perhaps at cost of sanity itself. 384pp. 5⅜ × 8½.
24669-8 Pa. $8.95

EGYPTIAN MAGIC, E. A. Wallis Budge. Sums up all that is known about magic in Ancient Egypt: the role of magic in controlling the gods, powerful amulets that warded off evil spirits, scarabs of immortality, use of wax images, formulas and spells, the secret name, much more. 253pp. 5⅜ × 8½.
22681-6 Pa. $4.95

THE DANCE OF SIVA, Ananda Coomaraswamy. Preeminent authority unfolds the vast metaphysic of India: the revelation of her art, conception of the universe, social organization, etc. 27 reproductions of art masterpieces. 192pp. 5⅜ × 8½.
24817-8 Pa. $6.95

CHRISTMAS CUSTOMS AND TRADITIONS, Clement A. Miles. Origin, evolution, significance of religious, secular practices. Caroling, gifts, yule logs, much more. Full, scholarly yet fascinating; non-sectarian. 400pp. 5⅜ × 8½.
23354-5 Pa. $7.95

THE HUMAN FIGURE IN MOTION, Eadweard Muybridge. More than 4,500 stopped-action photos, in action series, showing undraped men, women, children jumping, lying down, throwing, sitting, wrestling, carrying, etc. 390pp. 7⅞ × 10⅝.
20204-6 Cloth. $24.95

THE MAN WHO WAS THURSDAY, Gilbert Keith Chesterton. Witty, fast-paced novel about a club of anarchists in turn-of-the-century London. Brilliant social, religious, philosophical speculations. 128pp. 5⅜ × 8½.
25121-7 Pa. $3.95

A CÉZANNE SKETCHBOOK: Figures, Portraits, Landscapes and Still Lifes, Paul Cézanne. Great artist experiments with tonal effects, light, mass, other qualities in over 100 drawings. A revealing view of developing master painter, precursor of Cubism. 102 black-and-white illustrations. 144pp. 8¾ × 6⅜.
24790-2 Pa. $6.95

AN ENCYCLOPEDIA OF BATTLES: Accounts of Over 1,560 Battles from 1479 B.C. to the Present, David Eggenberger. Presents essential details of every major battle in recorded history, from the first battle of Megiddo in 1479 B.C. to Grenada in 1984. List of Battle Maps. New Appendix covering the years 1967–1984. Index. 99 illustrations. 544pp. 6½ × 9¼.
24913-1 Pa. $14.95

AN ETYMOLOGICAL DICTIONARY OF MODERN ENGLISH, Ernest Weekley. Richest, fullest work, by foremost British lexicographer. Detailed word histories. Inexhaustible. Total of 856pp. 6½ × 9¼.
21873-2, 21874-0 Pa., Two-vol. set $19.90

WEBSTER'S AMERICAN MILITARY BIOGRAPHIES, edited by Robert McHenry. Over 1,000 figures who shaped 3 centuries of American military history. Detailed biographies of Nathan Hale, Douglas MacArthur, Mary Hallaren, others. Chronologies of engagements, more. Introduction. Addenda. 1,033 entries in alphabetical order. xi + 548pp. 6½ × 9¼. (Available in U.S. only)
24758-9 Pa. $13.95

LIFE IN ANCIENT EGYPT, Adolf Erman. Detailed older account, with much not in more recent books: domestic life, religion, magic, medicine, commerce, and whatever else needed for complete picture. Many illustrations. 597pp. 5⅜ × 8½.
22632-8 Pa. $9.95

HISTORIC COSTUME IN PICTURES, Braun & Schneider. Over 1,450 costumed figures shown, covering a wide variety of peoples: kings, emperors, nobles, priests, servants, soldiers, scholars, townsfolk, peasants, merchants, courtiers, cavaliers, and more. 256pp. 8⅜ × 11¼.
23150-X Pa. $9.95

THE NOTEBOOKS OF LEONARDO DA VINCI, edited by J. P. Richter. Extracts from manuscripts reveal great genius; on painting, sculpture, anatomy, sciences, geography, etc. Both Italian and English. 186 ms. pages reproduced, plus 500 additional drawings, including studies for *Last Supper*, *Sforza* monument, etc. 860pp. 7⅞ × 10¾.
22572-0, 22573-9 Pa., Two-vol. set $35.90

THE ART NOUVEAU STYLE BOOK OF ALPHONSE MUCHA: All 72 Plates from "Documents Decoratifs" in Original Color, Alphonse Mucha. Rare copyright-free design portfolio by high priest of Art Nouveau. Jewelry, wallpaper, stained glass, furniture, figure studies, plant and animal motifs, etc. Only complete one-volume edition. 80pp. 9⅜ × 12¼. 24044-4 Pa. $10.95

ANIMALS: 1,419 Copyright-Free Illustrations of Mammals, Birds, Fish, Insects, Etc., edited by Jim Harter. Clear wood engravings present, in extremely lifelike poses, over 1,000 species of animals. One of the most extensive pictorial sourcebooks of its kind. Captions. Index. 284pp. 9 × 12. 23766-4 Pa. $10.95

OBELISTS FLY HIGH, C. Daly King. Masterpiece of American detective fiction, long out of print, involves murder on a 1935 transcontinental flight—"a very thrilling story"—NY Times. Unabridged and unaltered republication of the edition published by William Collins Sons & Co. Ltd., London, 1935. 288pp. 5⅜ × 8½. (Available in U.S. only) 25036-9 Pa. $5.95

VICTORIAN AND EDWARDIAN FASHION: A Photographic Survey, Alison Gernsheim. First fashion history completely illustrated by contemporary photographs. Full text plus 235 photos, 1840–1914, in which many celebrities appear. 240pp. 6½ × 9¼. 24205-6 Pa. $8.95

THE ART OF THE FRENCH ILLUSTRATED BOOK, 1700–1914, Gordon N. Ray. Over 630 superb book illustrations by Fragonard, Delacroix, Daumier, Doré, Grandville, Manet, Mucha, Steinlen, Toulouse-Lautrec and many others. Preface. Introduction. 633 halftones. Indices of artists, authors & titles, binders and provenances. Appendices. Bibliography. 608pp. 8⅜ × 11¼. 25086-5 Pa. $24.95

THE WONDERFUL WIZARD OF OZ, L. Frank Baum. Facsimile in full color of America's finest children's classic. 143 illustrations by W. W. Denslow. 267pp. 5⅜ × 8½. 20691-2 Pa. $7.95

FOLLOWING THE EQUATOR: A Journey Around the World, Mark Twain. Great writer's 1897 account of circumnavigating the globe by steamship. Ironic humor, keen observations, vivid and fascinating descriptions of exotic places. 197 illustrations. 720pp. 5⅜ × 8½. 26113-1 Pa. $15.95

THE FRIENDLY STARS, Martha Evans Martin & Donald Howard Menzel. Classic text marshalls the stars together in an engaging, nontechnical survey, presenting them as sources of beauty in night sky. 23 illustrations. Foreword. 2 star charts. Index. 147pp. 5⅜ × 8½. 21099-5 Pa. $3.95

FADS AND FALLACIES IN THE NAME OF SCIENCE, Martin Gardner. Fair, witty appraisal of cranks, quacks, and quackeries of science and pseudoscience: hollow earth, Velikovsky, orgone energy, Dianetics, flying saucers, Bridey Murphy, food and medical fads, etc. Revised, expanded In the Name of Science. "A very able and even-tempered presentation."—The New Yorker. 363pp. 5⅜ × 8. 20394-8 Pa. $6.95

ANCIENT EGYPT: Its Culture and History, J. E. Manchip White. From predynastics through Ptolemies: society, history, political structure, religion, daily life, literature, cultural heritage. 48 plates. 217pp. 5⅜ × 8½. 22548-8 Pa. $5.95

SIR HARRY HOTSPUR OF HUMBLETHWAITE, Anthony Trollope. Incisive, unconventional psychological study of a conflict between a wealthy baronet, his idealistic daughter, and their scapegrace cousin. The 1870 novel in its first inexpensive edition in years. 250pp. 5⅜ × 8½.   24953-0 Pa. $6.95

LASERS AND HOLOGRAPHY, Winston E. Kock. Sound introduction to burgeoning field, expanded (1981) for second edition. Wave patterns, coherence, lasers, diffraction, zone plates, properties of holograms, recent advances. 84 illustrations. 160pp. 5⅜ × 8¼. (Except in United Kingdom)   24041-X Pa. $4.95

INTRODUCTION TO ARTIFICIAL INTELLIGENCE: Second, Enlarged Edition, Philip C. Jackson, Jr. Comprehensive survey of artificial intelligence—the study of how machines (computers) can be made to act intelligently. Includes introductory and advanced material. Extensive notes updating the main text. 132 black-and-white illustrations. 512pp. 5⅜ × 8½.   24864-X Pa. $10.95

HISTORY OF INDIAN AND INDONESIAN ART, Ananda K. Coomaraswamy. Over 400 illustrations illuminate classic study of Indian art from earliest Harappa finds to early 20th century. Provides philosophical, religious and social insights. 304pp. 6⅜ × 9⅜.   25005-9 Pa. $11.95

THE GOLEM, Gustav Meyrink. Most famous supernatural novel in modern European literature, set in Ghetto of Old Prague around 1890. Compelling story of mystical experiences, strange transformations, profound terror. 13 black-and-white illustrations. 224pp. 5⅜ × 8½.   25025-3 Pa. $7.95

PICTORIAL ENCYCLOPEDIA OF HISTORIC ARCHITECTURAL PLANS, DETAILS AND ELEMENTS: With 1,880 Line Drawings of Arches, Domes, Doorways, Facades, Gables, Windows, etc., John Theodore Haneman. Sourcebook of inspiration for architects, designers, others. Bibliography. Captions. 141pp. 9 × 12.
24605-1 Pa. $8.95

BENCHLEY LOST AND FOUND, Robert Benchley. Finest humor from early 30s, about pet peeves, child psychologists, post office and others. Mostly unavailable elsewhere. 73 illustrations by Peter Arno and others. 183pp. 5⅜ × 8½.
22410-4 Pa. $4.95

ERTÉ GRAPHICS, Erté. Collection of striking color graphics: *Seasons, Alphabet, Numerals, Aces* and *Precious Stones*. 50 plates, including 4 on covers. 48pp. 9⅜ × 12¼.
23580-7 Pa. $7.95

THE JOURNAL OF HENRY D. THOREAU, edited by Bradford Torrey, F. H. Allen. Complete reprinting of 14 volumes, 1837–61, over two million words; the sourcebooks for *Walden*, etc. Definitive. All original sketches, plus 75 photographs. 1,804pp. 8½ × 12¼.   20312-3, 20313-1 Cloth., Two-vol. set $130.00

CASTLES: Their Construction and History, Sidney Toy. Traces castle development from ancient roots. Nearly 200 photographs and drawings illustrate moats, keeps, baileys, many other features. Caernarvon, Dover Castles, Hadrian's Wall, Tower of London, dozens more. 256pp. 5⅜ × 8¼.   24898-4 Pa. $7.95

AMERICAN CLIPPER SHIPS: 1833–1858, Octavius T. Howe & Frederick C. Matthews. Fully-illustrated, encyclopedic review of 352 clipper ships from the period of America's greatest maritime supremacy. Introduction. 109 halftones. 5 black-and-white line illustrations. Index. Total of 928pp. 5⅜ × 8½.
25115-2, 25116-0 Pa., Two-vol. set $21.90

TOWARDS A NEW ARCHITECTURE, Le Corbusier. Pioneering manifesto by great architect, near legendary founder of "International School." Technical and aesthetic theories, views on industry, economics, relation of form to function, "mass-production spirit," much more. Profusely illustrated. Unabridged translation of 13th French edition. Introduction by Frederick Etchells. 320pp. 6⅛ × 9¼. (Available in U.S. only)
25023-7 Pa. $8.95

THE BOOK OF KELLS, edited by Blanche Cirker. Inexpensive collection of 32 full-color, full-page plates from the greatest illuminated manuscript of the Middle Ages, painstakingly reproduced from rare facsimile edition. Publisher's Note. Captions. 32pp. 9⅜ × 12¼. (Available in U.S. only)
24345-1 Pa. $5.95

BEST SCIENCE FICTION STORIES OF H. G. WELLS, H. G. Wells. Full novel *The Invisible Man*, plus 17 short stories: "The Crystal Egg," "Aepyornis Island," "The Strange Orchid," etc. 303pp. 5⅜ × 8½. (Available in U.S. only)
21531-8 Pa. $6.95

AMERICAN SAILING SHIPS: Their Plans and History, Charles G. Davis. Photos, construction details of schooners, frigates, clippers, other sailcraft of 18th to early 20th centuries—plus entertaining discourse on design, rigging, nautical lore, much more. 137 black-and-white illustrations. 240pp. 6⅛ × 9¼.
24658-2 Pa. $6.95

ENTERTAINING MATHEMATICAL PUZZLES, Martin Gardner. Selection of author's favorite conundrums involving arithmetic, money, speed, etc., with lively commentary. Complete solutions. 112pp. 5⅜ × 8½.
25211-6 Pa. $3.95

THE WILL TO BELIEVE, HUMAN IMMORTALITY, William James. Two books bound together. Effect of irrational on logical, and arguments for human immortality. 402pp. 5⅜ × 8½.
20291-7 Pa. $8.95

THE HAUNTED MONASTERY and THE CHINESE MAZE MURDERS, Robert Van Gulik. 2 full novels by Van Gulik continue adventures of Judge Dee and his companions. An evil Taoist monastery, seemingly supernatural events; overgrown topiary maze that hides strange crimes. Set in 7th-century China. 27 illustrations. 328pp. 5⅜ × 8½.
23502-5 Pa. $6.95

CELEBRATED CASES OF JUDGE DEE (DEE GOONG AN), translated by Robert Van Gulik. Authentic 18th-century Chinese detective novel; Dee and associates solve three interlocked cases. Led to Van Gulik's own stories with same characters. Extensive introduction. 9 illustrations. 237pp. 5⅜ × 8½.
23337-5 Pa. $5.95

Prices subject to change without notice.

Available at your book dealer or write for free catalog to Dept. GI, Dover Publications, Inc., 31 East 2nd St., Mineola, N.Y. 11501. Dover publishes more than 400 books each year on science, elementary and advanced mathematics, biology, music, art, literary history, social sciences and other areas.